T0181445

# Digital Cultures

David Kergel

# Digital Cultures

Postmodern Media Education,
Subversive Diversity and Neoliberal
Subjectivation

 Springer

David Kergel
IU International University of Applied Sciences
Duisburg, Germany

Center for Diversity and Education in the Digital Age
Hildesheim, Germany

ISBN 978-3-658-35249-3        ISBN 978-3-658-35250-9   (eBook)
https://doi.org/10.1007/978-3-658-35250-9

This Springer imprint is published by the registered company Springer Fachmedien Wiesbaden GmbH, part of Springer Nature.
The registered company address is: Abraham-Lincoln-Str. 46, 65189 Wiesbaden, Germany

# Contents

# List of Figures

# Introduction

I wrote this book while traveling between New York and Duisburg, between Wikileaks and Google Maps. The main research interest on these journeys consisted of an attempt to interpret Internet culture or cultural practices in the digital age or the digital cultures.

As a heuristic starting point, I chose to ask about the spaces of self-determination and the dynamics of closure evoked by the digital cultures: How free and self-determined are we in the digital age? Does the Internet close us off in the sense of a digital society of control or does it open up new spaces for diversity and education?

The starting point of my analyses is the thesis that the Internet is both heritage and future: postmodern spaces of freedom and neoliberal definitions of the electronic age unfold in the ubiquitous cultural space that digital media span. At the same time, the Internet restructures social spaces in the material-physical world, digitalizes self/world relations and forms digital cultures, which in turn form us. In the process, the cultures of the digital exhibit a range between postmodern narratives of freedom and neoliberal self-optimization. They exist side by side and in opposition to each other and fight a battle over the symbolic order of the Internet. In the process, cultural lines of conflict from the material-physical world inscribe themselves on the Internet. These lines of conflict are a legacy of the protest movements of the 1960s and 1970s as well as the alternative culture of the 1980s and the increasing establishment of a neoliberal metanarrative since the 1980s. We as actors are embedded in these cultural interpretations offered by the Internet. The challenge is to gain a reflexive engagement with the complexity of the Internet as a space of cultural practices. In the sense of the difference-theoretical approach of hyperculture, the individual becomes the vanishing point of cultural practices in the digital age. This individuation-centeredness of the cultures of the digital results in the challenge of a critical-reflexive, self-determined sovereign handling of the ambivalence of the Internet between postmodern diversity and neoliberal subjectivization.

From the perspective of media education, the challenge arises to enable a sovereign, reflective approach to the cultural practices evoked by the Internet. This media pedagogical challenge can be analytically processed with the concept of media Bildung. Thus, a change of levels is carried out within the framework of this thesis. While the first three chapters are based primarily on methods of cultural theory and discourse analysis that draw integratively on media and epistemological analyses, the concluding fourth chapter takes a media education perspective. In this way, the cornerstones of postmodern media Bildung can be outlined. Such a postmodern media Bildung makes it possible to adopt a critical-reflexive relationship to the cultures of the digital, which were discussed in the first three chapters.

As with every book, I was again inspiring, supportive, loving, and indulgent accompanied by Birdy. Furthermore, I am indebted to Lennart and Merle for their support. I would also like to thank Ronald and all the scholars and students of the Department of Communication and Rhetorical Studies at Duquesne University, Pittsburgh, for their ideas, questions, and laughter.

IU International University of Applied Sciences                    David Kergel
Duisburg, Germany
Center for Diversity and Education in the Digital Age
Hildesheim, Germany

# Digital Cultures

**1**

## 1.1 Decentralized and Collaborative – The Genesis of the Internet

The 'Internet' is the name for networking computers or entire computer networks into a decentralized structure. This decentralized networking enables a 'many to many' communication instead of a 'one to many' communication, as it is, for example, inherent to television – the leading medium of the electronic age. Accordingly, Münker states that decentralization is the construction principle of the network (cf. Münker 2009, p. 51). Decentralization as the construction principle of the Internet "describes how the network is constructed – and how it works constructively. Not everything, but much of what is generated on or mediated through the Internet is an effect of decentralized moments" (Münker 2009, p. 51). This decentralized structure can also be seen in the development of the Internet. The Internet's growth and spread have always been linked to different social actors such as the military, researchers, entrepreneurs.[1] For example, Castells (2005) states that the Internet was "born primarily at the unlikely intersection of Big Science, military research, and a libertarian culture" (Castells 2005, p. 27).

The Advanced Research Products Agency (ARPA) was founded in 1958 on behalf of the U.S. Department of Defense. The reason was the so-called 'Sputnik

---

[1] When in the following we speak of 'the author', 'the researcher', 'the science', etc., this is an overgeneralization, since 'the scientist', 'the author', etc. do not exist as a generic singular. Nevertheless, this overgeneralization seems necessary in order to provide a conceptual approach to the concepts under discussion. As soon as reference is made to social categorizations or concrete actors, this is indicated accordingly.

© Springer Fachmedien Wiesbaden GmbH, part of Springer Nature 2023
D. Kergel, *Digital Cultures*, https://doi.org/10.1007/978-3-658-35250-9_1

Shock'. The so-called ARPA-Net was developed, in which (small) data packets could be sent back and forth by networking computers.

According to a traditional narrative that addresses how the development of the ARPA-Net came about, the political intention was to create an infrastructure for decentralized communication channels as well as command structures: even if (nuclear) attacks had wiped out central actors, communication and power to act should still be guaranteed. Kammenhuber et al. (2010) relegate this rationale to the realm of legends and underline that "during the attacks on September 11, 2001 [...] the Internet already reacted to localized failures of its infrastructure with network outages that sometimes lasted for hours" (Kammenhuber et al. 2010, p. 132). Accordingly, the power to act that the Internet was supposed to guarantee could not be realized, as the partial failure of the Internet in the context of the terrorist emergency of September 11 underlines (see Kammenhuber et al. 2010). In any case, it should be noted that the ARPA-Net represents an effect of military research:

> In the 1950s and 1960s, the Pentagon conducted and funded research of direct military use and a broad spectrum of basic and applied research for which the benefit to the military was not immediately apparent. The Advanced Research Projects Agency (ARPA) was set up to award such research contracts. Around 1966, the idea arose here to network computers working for ARPA projects since their computing capacity alone was often insufficient. It was expected that this would bring better results for the individual projects more quickly (Kirpal and Vogel 2006, p. 141).

Joseph Licklider, formerly a psychologist and computer scientist at the Massachusetts Institute of Technology (MIT), began focusing on computer interaction in 1962 as part of ARPA and recruited several scientists from leading universities such as Stanford University and the University of California, Berkeley.

The first connection between two mainframe computers was realized at universities in 1969. Initially, computers were connected between the University of California (UCLA), the Stanford Research Institute (RSI), and the University of Utah. The first message between these university computers was exchanged on October 29, 1969: At 10:30 p.m., a computer at the University of California sent the message 'lo' to a computer at the Stanford Research Institute, which was 600 km away.

The network grew steadily, and by 1971 it consisted of 14' nod. In 1972, the ARPA-Net was presented at the International Conference on Computer Communications in Washington, D.C., respectively, the form of decentralized data transmission was demonstrated (see Castells 2005, p. 20 f.). Since 1973, the focus has increasingly been on networking ARPA-Net with other computer networks. Standardized communication protocols have been developed for the spread of 'distributed computing' or for "interactiv[e] data transmission into decentralized computer network[s]" (Hartmann 2006, p. 171). These standardized communication protocols allow computers of different networks to communicate with each other, creating a network of networks – or the Internet. Since the late 1970s, more and more "interactive computer groups of scientists and hackers" (Castells 2005, p. 27) have developed decentralized computer networks. In the course of this networking, the use of e-mail, i.e., the sending of electronic messages, also showed how the media structure of the Internet promoted open communication: in 1971, computer technician Ray Tomlinson introduced an e-mail messaging system using the @ sign, which Palm (2006) sees as the "culmination of a cheap, egalitarian, democratic mode of communication" (Palm 2006, p. 125):

> As early as 1973, e-mail traffic accounted for about 75% of all traffic on the Net. Soon, e-mail was used to communicate about issues relating to the research topics being worked on and for an online discussion and clarify personal issues. In other words, the open, relatively unregulated exchange of information that is typical of the Internet emerged even in the early years of computer networking, on the one hand with the mode of operation that had occurred in the Network Working Group, and on the other hand by using the possibilities of the network beyond its actual purpose (Kirpal and Vogel 2006, p. 141).

At the end of the 1970s, several networks existed that needed to be interconnected. Specific technical solutions were developed for this purpose, such as the Transmission Control Protocol (TCP) and the Internet Protocol (I.P.), which enabled routing between network computers. The *Internet* Protocol also gave the *Internet* its name. In 1980, ARPA-Net was converted to the TCP/IP protocol, and in 1983, the U.S. Department of Defense declared it to be the standard for its computer networks[2]:

> Internet Protocol (I.P.) and the Transmission Control Protocol (TCP) [...] protocols define precisely how data is broken up into packets and sent between computers. Any computer using these methods (usually abbreviated to TCP/IP) should, in theory, be able to communicate with any other computer that also uses them. The Internet Protocol I.P. specifies the path that data takes between the connected computers or, as it is also called, the various "I.P. addresses." The TCP transmission protocol, in turn, ensures that the data packets are sent in a reliably orderly manner (Chatfield 2013, p. 5 f.).

---

[2] "Put simply, protocols are all the conventional rules and standards that govern relationships within networks. [...] In computer networks, scientists have formulated hundreds of protocols for years to govern e-mail, web pages, and so on, as well as many other standards for technologies that human eyes rarely behold" (Galloway and Thacker 2014, p. 291).

Also, in 1983, the "uneasy coexistence of military strategists and academic researchers, both of whom used the network, [...] led to its division into the MILNET (military) and the ARPRA-INTERNET (research), and in 1984 to the founding of the NSFNET" (Castells 2005, p. 32).[3] In the course of the first constitutional phase of the Internet, which can be located between 1970 and 1988, different semiotic understandings of the Internet gradually emerged:

> The military feared attacks on their central command and communication points, scientists wanted to expand their culture of autonomy, cooperation between peers, and the free exchange of information, and activists were looking for new fields of activity, and activists from the counterculture hoped-for new forms of free coexistence (Stalder 2016, p. 83 f.).

The Internet – analogous to its heterogeneous structure – was from the beginning open to interpretation and a place for different cultural practices or a projection surface for future possibilities of use. This is a characteristic that continues to shape Internet discourse to this day: In the course of increasing networking, "the Internet remained the preserve of a comparatively elite but globally dispersed circle. Primarily scientists and students exchanged their research results, discussed in online forums and newsgroups [...] or wrote e-mails to each other" (Kirpal and Vogel 2006, p. 142). With the beginning of the 1988/the 1990s, this network connection is further expanded, state and private networks are interconnected, so the Internet takes on global dimensions. This happens at a time of profound social transformations, which can be seen, among other things, in the collapse of the Soviet Union. In technological change, the ARPA-Net is abolished in 1990 and replaced by more modern forms of computer networking. Thus, at the Conseil Europèen pour la Recherche Nucléaire (CERN) in Geneva, Berners-Lee developed the foundations for the World Wide Web, which he published as software 1991. In doing so, the "key terms in Berners-Lee's design [...] were 'universal' and 'linked' or 'connected'" (Chatfield 2013, p. 8). To be able to realize the World Wide Web, nearly

> three components [used]: (1) the World's first web page with digital information (so that you could look at something), (2) a first browser program that allowed other users to see on their computer screens the information on that web page, and (3) the first web server, that is, a computer on which the web page was housed or "hosted." This

---

[3] In 1984, the MILNET or Military Network was created from parts of the ARPA-Net and was later replaced by the Non-classified Internet Protocol Router Network (NIPRNET), which is still in use today. The National Science Foundation Network, NSFNE for short, represented a computer network for scientific collaboration, to which Princeton University and Cornell University, among others, belonged.

host computer was supposed to function somewhat like a small digital notebook: The page with the information was stored on it, and then anyone could easily connect to it via browser and view the page. No matter how many browsers accessed this page simultaneously, it remained on the host computer (Chatfield 2013, p. 8 f.).

Berners-Lee programmed the so-called hypertext protocol or Hypertext Transfer Protocol (HTTP). This protocol enables the computer to search for files on the Internet to link them to a document – "HTTP is a protocol for transferring information with the efficiency necessary for making hypertext jumps. [...] HTTP is, therefore, a simple request/response protocol" (Berner-Lee et al. 1994, p. 794). Through HTTP, a hypertext structure and thus a hypertextual connection between computers can be realized. In this process, the digital text is marked with a small number of 'tags' embedded in the text's electronic form. Each tag contains detailed information that tells a browser how to display that part of the text and how other pages somewhere on the web should be linked (cf. Chatfield 2013, p. 10). The documents can be organized in multimedia, contain videos, images, text, and be hyperlinked to other documents. The hyperlinks allow these linked documents to be accessed from the form currently present. In doing so, the files have been given a specific Internet address – the so-called Uniform Resource Locator (URL). Berner-Lee christened the program that allowed documents transmitted over the Internet to be presented on a browser. "Browsers allow us to navigate between different websites and web pages by either following links or typing in a specific address" (Chatfield 2013, p. 10). Graphics-enabled browsers make navigating the Internet significantly easier. The markup language (or machine-readable language) Hypertext Markup Language (HTML) enables data to be structured so that the browser can display it: "HTML is defined to be a language of communication which flows over the network. There is no requirement that files are stored in HTML" (Berner-Lee et al. 1994, p. 795). Berner-Lee coined the term World Wide Web or W3 for this networking or this form of data transfer – "He reflected the idea of the practice worldwide linking of files and the possibility of combining them, again and again, using hypertext in the term World Wide Web, literally worldwide web" (Kirpal and Vogel 2006, p. 143).

Gradually, the Internet became a mass-market from 1989/1990 onwards. "Throughout the 1990s, the number of Internet users doubled on average every year, a growth rate that only diminished slightly in the following decade" (Chatfield 2013, p. 6). In 1990, the National Science Foundation decided to open the Internet to commercial uses. With the establishment of the Integrated Services Digital Network system (ISDN), data could be sent without much delay.

After initial pilot projects, ISDN technology was put into practical service in 1989. This technology and the commercial opening led to an 'explosive development' (Castells 2005, p. 19) and the Internet's spread.

In 1990, the first commercial Internet provider, World, entered the market. In 1994, Netscape Communications made available the first commercial browser, Netscape Navigator. These inventions finally paved the way for the commercialization of the Internet and its cultural opening: in 1994, the number of commercial Internet users exceeded that of scientific users. In 1996, some 16 million computers were part of the network. In 1998 – the year Google was founded – there were about 36 million computers. The rapid spread of commercial use of the Internet was also an effect because more and more households had a computer.

Commercialization led to a paradigm shift in the orientation of the Internet: "The ambition to make money with the WWW was not what the inventors had in mind, but it could not be stopped in the longer term. Over time, the Web assumed a size that almost inevitably made it attractive to commercial considerations" (Kirpal and Vogel 2006, p. 144). Castells notes that the Internet was privatized until the mid-1990s (see Castells 2005, p. 26). For example, Amazon started as an online retailer in 1995. Products could be marketed online and sent directly from the manufacturer to the customer. A "virtually unlimited growth market" (Kirpal and Vogel 2006, p. 145) seemed to emerge. "The commercial potential of the new technology seemed limitless" (Hartmann 2006. p. 181). The privatization and commercialization of the Internet affected the first economic crisis: the New Economy was discursively exaggerated, which led to "[v]irtual investors [...] rushing to secure entry into the unknown market" (Hartmann 2006, p. 181). Among other things, "company takeovers, countless start-ups and IPOs between 1995 and 2000" (ibid.) created a speculative bubble that led to the so-called dot-com crash.

Despite these slumps, the Internet continued to develop steadily with Wikipedia being published in 2001, Facebook going online in 2003, and YouTube in 2005. In 2003, the term Web 2.0 was coined as a decisive concept in the discourse on the cultural positioning of the Internet. After the commercial opening of the Internet, the establishment of Web 2.0 represents another major paradigm shift in the culture of the Internet. With Web 2.0, the Internet's subsequent development in the electronic age led to a digital age. With the digital age new forms of self/world relations emerged which unfolded in digital cultures. These digital cultures also manifest in the discourses that are to be located in the context of the term Web 2.0.

Web 2.0 empowers the user and constitutes the user's need to unfold and thereby. In consequence, Web 2.0 oscillates discursively between postmodern education and neoliberal subjectivation. *Against the background of Web 2.0, digital cultures, i.e., specific forms of using the Internet and digitally shaped self/world relations in social contexts, have in common that they are characterized by a high degree of participation in the discursive thematization.* This empowerment of the user is a significant feature of the digital. This thesis will be developed in the following.

## 1.2   A 'Grid of Numbers' or the 'Digital Folding' of Reality

The Internet enables decentralized many-to-many communication. The development of the Internet was a collaborative process in which decisive developments resulted from the communicative openness of the Internet. For example, the development of electronic mail or e-mail was less programmatically targeted but rather an effect of users' behavior. The development of the e-mail represents paradigmatically the openness and spontaneity that characterizes the Internet as a communication medium. Decentralized communication and the establishment of Web 2.0 with its poly-directional and polyphonic potentials at the beginning of the 2000s are part of a social change. This social change leads the civic society cultural discourses from the electronic age into the digital age. The electronic age was characterized by a 'de-familiarization' (see Han 2005) of time and space. This change can be exemplified by television – the leading medium of the electronic age. Television

> has dominated Western popular culture and general communication relations from the middle of the 20th century until around the millennium turn. It has taken us from the age of classical, analog mass media such as newspapers, film, and radio to the age of digital and networked media (Engell 2012, p. 12).

Through television, images from all over the World reach us. This is an effect of the electronic age: Distances erode, the telegraph, and later the telephone 'remove' communication. Through this implosion of time and space, the world shrinks into a 'global village' (see McLuhan 1968).

This implosion of space is being decentralized by the Internet and its many-to-many communication structure. The Internet's communication potentials were developed in the context of the electronic age and further unfolded in the digital age. The decentralization of communication, the mass distribution of the Internet and its commercialization, the development of participatory Web 2.0 tools such as wikis, weblogs, and podcasts led to a new digital-based communication culture. From this perspective, the Internet is constitutively performative: without interaction, the Internet 'dies'. This interactive dimension is also evident in the digital structure, which constitutively helps define the Internet's communication processes and is a central feature of digital cultures. This production- and action-oriented interactive dimension of the digital will be further elaborated in the following.

Digital media constitute the Internet by storing, transporting as well as representing data – "A computer requires that everything from the continuous flow of our everyday reality is converted into a grid of numbers that can be stored as a representation of reality that can then be manipulated with algorithms" (Berry 2014, p. 48). Communication processes via the Internet are digitally based and require transfer performance through digital codes which constitute a 'grid of numbers':

> This minimal transformation is influenced by a socio-technical device's input mechanism in which a model or image is stabilized and perceived. It is then transformed internally, depending on a series of interventions, processes, or filters, and finally rendered as a final result, usually in visual form (Berry 2014, p. 48).

Unlike the media of the book culture (books, newspapers) or the electronic age (radio, television), digital media are defined by having the computer as their technical core. Computational processes mediate media content, and, in the course of this, the content is restructured in a binary form (see Zorn 2011, p. 176). The representation of media content is the result of calculations. As computer processes, these calculations represent automated activities that have been written by programmers:

> To let the computer do something, processes of activities have to be translated into describable models. For this purpose, actions must be recorded and formally described to be operationalized and broken down into small parts. These individual steps are translated into characters, into programming language [...], so that the computer can execute them through programming (Zorn 2011, p. 177).

One can speak about a 'digital principle' which represents "a discrete number representation in the binary form" (Hartmann 2006, p. 185). This digital principle "has proved to be a technically proven means of electronically processing and storing data in computers" (ibid.). In this process, data is "decomposed into a series of two-valued states that can be automatically processed as switching states" (ibid.). The digital code makes it possible to make information technically processable. These translation and transmission processes inscribe in the media content – or constitute it. Thus, digital media as information-mediating instances are involved in the "production of media content" (Schelhowe 2007, p. 46). From a media theoretical perspective, Schelhowe points out that digital media "themselves produce the message" or are "involved in this production in a quite fundamental sense" (ibid.). In terms of a content-form relation, the media content provided cannot be separated from the technical side of the media – "information in digital media is not simply digitized, input and then output again in equal measure as media content, but it is processed in these operations by computational processes and thereby changed" (Zorn 2011, p. 179). From this perspective, an ephemeral structure of digital media content emerges. Digital media contents are not fixed objects, but are "created anew in their manifestation in the course of the interactive communication structure of the Internet" (Schelhowe 2007, p. 47). They exist "in memory as digital objects that are only realized in their respective different manifestations through the process, the program" (Schelhowe 2007, p. 47). Digital media objects have the potential to be 'different every time.' Many different instances can be created from the 'atomized data' available from one program (see Schelhowe 2007, p. 47). From this perspective, digital media content "are never finished objects. They remain 'unfinished,' are process rather than a product" (ibid.). In line with the ephemeral structure of digital content, time is restructured as a constitutive communication factor. Thus, "with instantaneous information transmission [...] everything that is not 'here' is unreachable and everything that is not 'now' has disappeared" (Stalder 2016, p. 147). *The present tense becomes a feature of media content:* "The spatio-temporal horizon of digital communication is global, that is, placeless permanent present" (ibid.). Thus, Raulet (1988) points out that

> [n]umeric images [...] no longer [know] before and after; time ceases, as the a priori frame of sensuality, as Kant called it, to create that continuity of experience that is constitutive for the reality of the cognized. Images follow one another as arbitrary moments, as interchangeable snapshots that no longer obey any temporal hierarchy (Raulet 1988, p. 169).

Through the Internet, the content to be communicated in information exchange becomes temporally immediate, an aspect that can be summarized in media theory with the implosion of time *through* the Internet or *on* the Internet. This implosion of time is illustrated by Wandtke (2001) regarding the communication possibilities of websites:

> The website, unlike the film, is produced in a virtual space. The website is inconceivable without the Internet. The film work can exist without the Internet. With the Internet, a new element has been introduced into intellectual production, as it were. Besides, the website is not only produced by the author using a computer into the Internet, but the traditional process of division of labor between production, distribution, circulation, and consumption is, as it were, suspended in time (Wandtke 2001, p. 21).

Increasing "digitization, which makes all content editable, and networking, which creates an almost endless mass of content as 'raw material'" (Stalder 2016, p. 67), leads to a restructuring of the media perception of reality. This also changes the form of reception of media content: Digital media represent an invitation to interact, "a challenge to intervene in the process oneself, to shape it" (Iske and Marotzki 2010, p. 47). The knowledge that emerges, for example, in the context of Web 2.0 applications such as wikis and blogs can be "characterized as a result of interactions" (Iske and Marotzki 2010, p. 142). The digital dimension of the Internet requires interaction that is characterized by a performative process-orientation and incompleteness and relies on the participatory involvement of users – "Due to the basal logic of user-generated content, the Internet user has a part in the generation of discourse and knowledge landscapes. Thus, he is from the outset part of a larger community that interacts" (Iske and Marotzki 2010, p. 142). The "digital 'folding' of reality" (Berry 2014, p. 47) changes everyday practices, leading to a "multiplication of cultural possibilities" (Stalder 2016, p. 10) or cultures of the digital. These range from anarchist subversion to digital-based neoliberal (self-)optimization but are always defined by participation as a premise.

Actors performatively generate self/world relations through their use of digital media. *Media change inscribes itself in cultural practices and redefines the way culture is theorized. The 'digital folding of reality' also manifests in reflecting the medial in cultural theoretical discourses.*

## 1.3    Culture in Medial Change

### 1.3.1    Culture – An Attempt at Definition

The concept of culture can be understood as a concept in flux: With societal self-understanding discourses, the respective understanding of culture also changes. The change in the interpretation of culture can be exemplified by the current social change in national culture: As a result of the so-called European 'refugee crises', concepts such as national culture, leading culture, and integration are once again becoming discursively virulent. The demand for integration, in turn, presupposes the question of what is to be integrated. Only the answer to *what* should be integrated into allows the question of *how* to be raised. An implicit starting point for these discussions is how cultural, social self-image should serve as a normative reference for integration strategies.

The increasing thematization of national identity points to the increased discursive relevance of the concept of culture. This is contrasted by the difficulty of framing the term culture in scientific terms. The attempt to provide a final definition of culture already seems to fail because understanding culture is tied to the respective discourses of societal self-understanding discourses. This discursive conditionality of the interpretation of the notion of culture is exemplified when Hahn (2013) traces how the concept of culture changed in the field of ethnology according to in the course of social dynamics: in the age of imperial colonization, the culture of the colonial power is seen as superior to the cultural identity of colonized peoples (see in the sense of a primary source Taylor 1871). On the other hand, in a postmodern social self-thematization, the conditionality of one's cultural identity and one's cultural claims to truth are critically reflected. Thus, with social transformations (e.g., from modernity to postmodernity), what is understood by culture also changes (see also Luhmann 1995): "Cultures are not scientific 'objects' (assuming such things exist, even in the natural sciences). Culture, and our views of 'it,' are produced historically, and are actively contested" (Clifford 1986, p. 18).

In the sense of a working concept, a concept of culture is outlined below. This concept follows a totality-oriented understanding of culture. In this context, culture is understood as the totality of thought patterns, actions, and perceptions of groups/collectives. From this perspective, culture represents an ensemble of collectively disseminated/shared forms of belief, knowledge, and life, which actors acquire through socialization. This approach holistically encompasses cognitive, emotional, and non-reflective forms of experience and practices of action. Such a concept of culture can be constructed based on an etymological reference: The word culture derives from Latin (cultura) and denotes 'cultivation,' 'care,' 'farming.' Regarding this etymological root, everything that man brings forth in a productive way can be defined as culture in the broadest sense. From this perspective, culture is to be distinguished from nature, which is not created by man. If man changes nature, he cultivates it. Against the background of these considerations, culture can be understood as a collective term to utilize the environment. This utilization of the environment ranges from the development of technical instruments to aesthetic-semiotic processes of meaning formation. This utilization of the environment appears in a 'group-based form: "'Culture' here refers to the extension of that group (or society or civilization) for which the cultural contents or practices in question are characteristic" (Welsch 2010, p. 1). This group-based dimension of the utilization of nature constitutes cultural identity in the form of a group membership.

Culture thereby includes a relation of difference to other groups. From this perspective, culture is constitutively an experience of difference.[4]

Culture has a material sphere, as can be seen from Heidegger's understanding of technology – "technology is a human activity" (Heidegger 2006, p. 6). In his engagement with the World, man creates the World he appropriates, or instead, he 'brings it forth.' "Her-vor-bringen [bringing-forth] occurs only insofar as the hidden comes into the unconcealed. This coming is based and resonates in what we call unconcealing" (Heidegger 2006, p. 11). From this perspective, culture is a process of 'unconcealing' or the bringing forth of world by human beings. For example, "that which is brought forth by craftsmanship and art [...] has the dawn of bringing forth not in itself, but [...] in the craftsman and artist" (Heidegger 2006, p. 11). Culture is brought forth by transforming what is not present into something present in the world's engagement. Accordingly, Heidegger refers to agriculture as a form of uncovering – the Latin name representing the etymological root of culture.

*If technology denotes the productive engagement with the World, it can be understood as an act of culture. From this perspective, dealing with technology is an action that constitutes culture.* The technical dimension of culture thus comes into focus. From the "co-constitution of technology and being human" (Fabretti 2014, p. 101) emerges culture. 'Digital technologies' (see Fabretti 2014, p. 89) are not primarily an instrumental relationship to exchange information via Facebook. Instead, being human in the digital age is constituted *through* digital media and (self-)narratives *via* digital media. Such a performative understanding of culture implies that the technology with which culture is produced also changes with the medial change. Technology as a process of action that constitutes culture consequently also always has a medial dimension. The conceptual version of culture changes not only in the course of the change of societal self-understanding processes but also in the medial change, through which new forms of 'bringing forth' arise. It may be concluded that an inseparable interlocking of societal self-understanding discourses with technical and medial change can be assumed. In this infinite change, culture and cultural understandings are continuously redefined.

In the following, this thesis will be analytically elaborated based on the postmodern cultural concepts of trans-culture and hyper-culture.

To begin with, Herder's concept of culture will be presented. Herder's concept of culture is brought forward because Herder's understanding of culture often serves as a discursive point of orientation for the conceptual discussion of culture –

---

[4] If one follows the argumentation of understanding culture as a 'group-bound utilization of the environment,' different 'utilization strategies of the environment' meet in intercultural encounters.

for example, in Welsch's (2010) modeling of trans-culture. Following Herder's concept of culture. Simmel's concept of the stranger is outlined. This concept can serve as a heuristic strategy for analyzing culture as a construct of difference: How is the relation between cultures defined?

As mentioned above, culture is constitutively characterized by a relation of difference to other cultures. Simmel's concept of 'the stranger' can be used to work out how the relationship between cultures is conceived as a relationship of difference between the self and the stranger in the respective understandings of culture.

## 1.3.2   On the Blissful People and 'the Prejudice Is Good'

The reference to Herder provides a fundamental point of orientation for the conceptual discussion of culture. Accordingly, in the "center of historical argumentations [...] Herder appears again and again with many authors" (Neubert et al. 2013, p. 26). One of the reasons is that "Herder's philosophy [...] represents an essential starting point for cultural-critical thinking in modern times" (ibid.). According to Herder, culture is defined out of a specific historical situation. The formation of culture depends on what, among other things, 'God, climate, time and stage of the world age' (see Herder 2012, p. 32) have made of people – From this perspective, people appear as the representatives of culture. Due to a people's specific situation, a culture-specific form of the development or 'satisfaction' of the (cultural) 'needs' results. The specific situation of the respective people leads to different cultures and different cultural self-understandings. The range of different cultures makes cultures and thus peoples *not* comparable with each other – at least in an evaluative sense:

> As soon as the inner sense of happiness, the inclination, has changed: as soon as the outer opportunities and needs form and strengthen the other sense, who can compare different senses' different satisfaction in different worlds? The shepherd and father of the Orient, the husbandman, and artist, the sailor and competitor, conqueror of the World – who can compare? In the laurel wreath or in the sight of the blessed herd, in the merchandise ship and captured field sign lies nothing – but in the soul that needed that, strove for that, has now attained that, and wanted to attain nothing but that – every nation has its center of happiness in itself, as every sphere has its center of gravity (Herder 2012, p. 34).

Herder introduces an ontological separation between cultures. To justify this, Herder transfers the concept of perfect happiness/bliss formulated by Aristotle in the Nicomachean Ethics to the formation of nations: Happiness represents a self-relationship defined by an individual realizing a balanced relationship between his needs. Virtue, of which bliss represents the highest, is 'a *habitus of choosing, which holds the mean measured after us, and is determined by reason, and as a wise man is wont to determine it'* (Aristotle NE 1106 b35–1107 a). It is necessary to find a state of 'well-being' between poles such as 'intemperance' and 'dullness' or the poles of 'cowardice' and 'foolhardiness.' Each individual has to realize this middle ground for himself, to balance it out again and again. Consequently, finding one's middle represents a performative, infinite process of self-localization. By transferring this concept of happiness to the people and thereby defining the people as bearers of culture, Herder introduces a prescriptive dimension into the discussion of culture: Every person must develop their cultural potentials to find its center *versa*. This theoretical move by Herder transforms the discussion of culture from descriptive analysis to ethical discourse. Thus, Kristeva (1990) points out that the folk spirit "is not biological, 'scientific' or even political [in Herder], it is moral by its very nature" (Kristeva 1990, p. 192). Here, this ethical dimension does not pose the question of an encounter of cultures and does not aim at an 'inter' of cultures. Rather, cultural encounters are to be understood more as processes of demarcation, which Herder describes in condensed form in the so-called 'sphere model':

> The Greek makes his own as much of the Egyptian, the Roman of the Greek, as he needs for himself: he is sated, the rest falls to the ground, and he does not strive for it! Or if, in this formation of his national inclinations into his national happiness, the distance between people and people has already grown too far: behold how the Egyptian hates the shepherd, the tramp! How he despises the careless Greek! Thus each of two nations, whose inclinations and circles of happiness clash – it is called prejudice! Scurrilousness! Restricted nationalism. Prejudice is good, in its time: for it makes happy. It presses peoples together to their center, makes them firmer on their trunk, more flourishing in their kind, more ardent, and thus also more blissful in their inclinations and purposes (Herder 2012, p. 36).

Herder's argumentation shows analogies to epistemological strategies based on the theory of difference: the cultural self-understanding of a people is constituted in the process of demarcation. The discursive (hostile) occupation of other cultures simultaneously leads to a – at least subtextual – revaluation of one's own culture. This own culture does not exhibit the 'negative characteristics' identified in other cultures. In the course of this cultural demarcation process, one's characteristics can be more strongly identified as positive reference points and thus help to constitute one's cultural self-image.[5] In this demarcation process, the people are

---

[5] Cultures develop quasi-evolutionarily (see Kristeva 1990, p. 192), so that a world history of humanity is a world history of cultures (see Herder 2012, p. 109 f.) – an approach that Hegel, among others, takes up and expands in his "Lectures on the History of Philosophy" ("Vorlesungen der Geschichte der Philosophie"; 1805–1836).

constituted as the representants of a (cultural) nation. Herder thus prefigures the concept of the cultural nation. This concept undergoes a discursive politicization in the context of the defeat of Frederick the Great's army by the French Revolutionary troops: "Only after 1806 does this *cultural* concept of 'nation' become a *political one*, deployed in the national-political struggle" (Kristeva 1990, p. 192). In his work "Ideas on the Philosophy of the History of Mankind" ("Ideen zur Philosophie der Geschichte der Menschheit"; 1784–1791) Herder defines the nation as a representant of culture; in the course of its self-development, a country constitutes as nation and also develops its language (see Herder 2016, p. 185). This perspective on language also reveals the medial implications of Herder's conception of culture: culture is shaped by language as central medium of communication. In other words: Herder's reflection on culture has a constitutive medial dimension. Cultures manifest themselves metonymically in language or in the way a people use language. Herder anticipates the central consideration of the linguistic turn, that language represents a medium of knowledge: Only about that which has a signification in the cognitive medium of language can one speak 'only in a language, can have a meaning.' (see Frank 1984, p. 283). In the sense of the linguistic turn, social reality or social knowledge is produced through language. Herder assumes that man is a creature of language (see Hartmann 2000, p. 83): "All of us come to reason only through language" (Herder 2016, p. 184). Language metonymically manifests the self/world relationship of a people and thus represents the central metonymy of cultural identity. If the culture changes, the language changes (see Herder 2016, p. 185). Since language is a central manifestation of cultural identity, individuals from other cultures can never adequately learn the language of another culture:

> Therefore, because people's language, especially in books, is shy and fine, not everyone is fine and shy who reads these books and speaks this language. How he reads them, how he speaks them, that would be the question; and even then, he would always repeating: he follows the thoughts and the power of designation of another. (Herder 2016, p. 188).

According Herder, language draws an ontological division between peoples. This consideration also has consequences for the relationship between individual and culture: the individual represents a culture whose language he or she speaks. Hartmann (2000) points out that one of Herder's objectives was "to re-educate the Germans in a lost poetic mother tongue, i.e., to treat the nation through language" (Hartmann 2000, p. 83). Language thereby "clearly takes the place of religion as a socially unifying force" (ibid.). In doing so, language unites through the experience of sound, an aesthetic experience, and metaphysical reason and represents medial "the sensual and the ideal" (Hartmann 2000, p. 83). With the thematization of language, a medial dimension of culture becomes subtextually relevant: Cultural identity unfolds through language. This medial dimension of cultural identity becomes even more clear, when Herder links the development of language to printing technology, which was relatively new at the time: language writes itself into books. Printing enables a collection of knowledge in written language:

> At last, even – marvelous invention! – mémoirs and dictionaries, where everyone can read what and how much he wants – and the most glorious of glorious inventions, the dictionary, the encyclopedia of all sciences and arts [...] What the art of printing has become to the sciences, the encyclopedia has become to the art of printing: the highest summit of expansion, completeness and eternal preservation (Herder 2012, p. 77).

Books provide an access to other cultures, which is the "practical[e] side of the book" (Herder 2012, p. 64). Even if other cultures' works can only be inadequately depicted in books and the language of other people can never be adequately understood due to cultural barriers (see Herder 2012, p. 64 f.), the book remains central access to other cultures.[6] The book becomes the guiding medium of knowledge as it materializes language, which is the metonymy of reason. "The book, this author, this set of books is to educate" (Herder 2012, p. 65). In his analysis of the book-culture, McLuhan elaborates that written language fostered specific cultural effects. For example, standardization and systematization of written language led to a national language. The "dynamic logic of printing" (McLuhan 1968, p. 312) unfolded a "centralizing unifying force" (ibid.) and thus brought about a national language in the first place. At the same time, differences are manifested in written language: language marks the ontological difference between cultures. Language is either culturally own or culturally foreign. Derived from the cultural dimension of language, consequences can be drawn for cultural analysis in general: Definitions of culture always include the implication that different cultures exist that are to be delimited from one another. From the perspective of cultural theory, it is necessary to determine the constellations of demarcation between what is culturally one's own and what is culturally foreign. Here, we can refer back to Simmel's foreign concept, which makes it possible to take an analytical look at the cultural dynamics of demarcation outlined by Herder.

---

[6]This book-bound perspective on other cultures found its application in the so-called armchair anthropologists, who knew foreign cultures from books rather than from an encounter on the ground (see Sera-Shriar 2014).

### 1.3.3   The Opening of Culture – Simmel's Concept of the Stranger

Herder develops an approach to cultural theory that is defined by the fact that cultures constitute a self-understanding through demarcation dynamics from other cultures. This demarcation process leads to the so-called sphere model, which can be interpreted so that cultures constitute their center from relations of difference to other cultures – constellations of exclusion-inclusion constitute culture. From this perspective, a culture can only exist in a superordinate context of cultures and requires actors who belong to the cultural community and actors who do not belong or who are foreign. The exclusion-inclusion logic that characterizes Herder's understanding of culture is conceptually elaborated by the sociologist Simmel in the 1908 text "The Stranger." In the context of this impactful text (see Bodeman 2011), which forms a chapter of the book "Sociology. Inquiry into the Forms of Socialization" ("Soziologie. Untersuchung über die Formen der Vergesellschaftung"), Simmel conceptualizes the stranger as a social type that represents the foreign. Simmel's approach influenced, among others, "his most important contemporaries in sociology: Max Weber, Werner Sombart, Ferdinand Tönnies, and Robert Michels. Later, the Chicago School, especially Robert Park and Everett Hughes, transposed the idea of the stranger as a 'marginal man' into American sociology" (Bodeman 2011, p. 185).

According to Simmel, the foreign is constituted in relation to the own – analogous to Herder, who describes cultures as effects of a differential structure. In the relational structure between the own and the foreign, the foreigner is defined by *potential* mobility. "The stranger is meant here, then, not in a sense hitherto often touched upon, as the wanderer who comes today and goes tomorrow, but as the one who comes today and stays tomorrow" (Simmel 1908, p. 1). He is a constitutive part of the group to which he is a stranger since his otherness marks the group's boundary. The stranger is consequently "fixed within a certain spatial circumference – or one whose boundary determination is analogous to the spatial one" (ibid.). At the same time, the stranger also signals something beyond the group. It constitutes the group and transcends it at the same time. The demarcation from the foreign constitutes the 'own.' This formulates a basic principle of differential identity construction. As the known unknown, the existence of the foreign constitutes in this process the consciousness of one's own, familiar, non-foreign culture. This dynamic of the interplay of differences between the self and the foreign can also be made useful for cultural analyses. In the interplay between being included in a cultural space into which he does not belong, the stranger constitutes a "unity of nearness and remoteness" (Simmel 1908, p. 1). The stranger represents the culturally other and, at the same time, brings the "distant near" (ibid.). The stranger is "an element of the group itself [...] an element whose immanent and member status at the same time includes an outside and opposite" (ibid.). Here Simmel goes a decisive step further than Herder since cultures always have an opening through the stranger, despite demarcations from the stranger. The stranger is constitutive of the self and opens the self to foreign influences. The stranger comes from the outside and establishes a mediating function to the outside. In terms of media theory, the stranger as a mediator is a metaphorical figure for the medium as mediating element. In his mediating function, the stranger becomes a trader: he delivers goods that are needed and come from abroad. From this perspective, one's 'own' is defined by a certain unfamiliarity: The foreign is not the unknown. What is not known to one cannot be foreign to one. The foreign is consequently a relational concept. It captures the relation between the poles of near and far. From this perspective, the stranger is consequently "beyond far and near" in a constant 'inter.' The stranger is part of a liminal space. The stranger's otherness is defined, among other things, by the fact that he has no ground. "The stranger, by his very nature, is not a landowner" (Simmel 1908, p. 2). The metaphor of taking root indicates that the stranger appropriates soil and it integrates. By taking root, essential characteristics of one's own are adopted. The foreigner alienates himself by becoming culturally one's own – in abstract terms, the constellation of integration is described here.

Another possible interpretation of Simmel's figure of the stranger lies in its objectifying function. Instead of integrating, the stranger as a dialogue partner can objectify the own: the stranger brings another non-own perspective into the discourse: he is objective since he has not been socialized into the values and traditions of the own.[7] Through his liminal relationship between proximity and distance, whose dialectic he represents as the stranger's social type, the stranger can also be used as a negative referent. It becomes possible to stage the stranger discursively as a projection surface for the danger to the stability of the self's symbolic order. The foreigner attacks and threatens one's own: "From time immemorial, in uprisings of all kinds, the party under attack has claimed that there has been an agitation from outside, by foreign emissaries and agitators" (Simmel 1908, p. 2). From this perspective, the stranger moves close to Freud's approach to the

---

[7] Similar considerations underlie Brecht's alienation effect: plays that are committed to the approach of epic theatre are intended to provide the recipient with an alien perspective on the familiar. In this way, an objective, analytical attitude to social processes can be adopted. Social practice can be reflected and, according to Brecht, changed in the sense of Marxist approaches.

uncanny. The uncanny alters the homely, subverts the familiar, reverses the homely. The stranger can thus be understood as the bearer of the uncanny: The familiar own is lost through the stranger's presence – stranger and own are dichotomously opposed in this constellation. The constellation of the uncanny is often actualized in (right-wing populist) discourses on cultural identity: In the context of right-wing populist cultural demarcations, the foreign is discursively thematized as a sinister intruder that threatens one's own culture. Paradigmatically, such a discursive thematization of the foreigner as a negative referent can be stated in the case of the political group chairman of the right-wing populist party of the AFD (Alternative for Germany, Alternative für Deutschland = AfD), Alexander Gauland. As an example, an excerpt from an argument published in Cicero in July 2016 between Gauland and the FDP politician Wolfgang Kubicki is cited below:

> We are a party of people who are concerned who are afraid of foreign infiltration. They don't want a million foreigners traveling around in this country who are not politically persecuted at all. These are small people who want to keep their Germany a little as it once was (Gauland in Cicero July 2016, p. 19).

Gauland actualizes the topoi of the foreign and the own. The foreigner is accompanied by the threat scenario of "foreign infiltration." Foreign infiltration represents a threat to the "little people." The specification by the attribute "little" reinforces the discursively manufactured threat of foreign infiltration (not 'big,' defensible people are named, but "little"). On the other hand, the foreigners are determined only quantitatively and through their travel and threaten the stability of 'their own' Germany ("their Germany") through their numbers and their mobility. This is a paradigmatic example of how Simmel's model of the stranger can be used as a heuristic analytical strategy to reconstruct exclusion-inclusion constellations in cultural discourse: Gauland closes the cultural space and ignores the fact that the foreign is constitutively part of the own and only constitutes this own via demarcation from the other. Consequently, the own is never really closed but constitutively open. The stranger, as a wanderer from afar, who remains and yet signals the mobility of the potentially wandering, who comes from abroad, opens up spaces. In doing so, the stranger mediates between the own and the foreign, which can be seen in Simmel's metaphor of the merchant. From the perspective of cultural theory, the consequence is to conceive of cultures in permissive terms. This erodes a dichotomous contrast between the foreign and the own, as Gauland has discursively actualized it.

It is possible to apply Simmel's concept of the foreign to media theory model and relate it to the Internet as a new and thus foreign medium. From this perspective, the Internet appears as a new medial dimension of reality experience or as the foreign, which evokes irritation and demarcation strategies. This thesis will be developed in more detail in the following.

### 1.3.4   The Internet as the Culturally Alien

From a media theoretical perspective, the stranger can be applied to the Internet as a heuristic analysis strategy. The Internet opens up spaces and thus symbolizes the presence of the foreign. From this perspective, the Internet "as the basic technology of unbounded *cyberspace* competes with concepts of the territorial and the national" (Hartmann 2006, p. 165, ed.). The Internet connects computers to networks and distances erode. The foreign enters the sphere via the Internet through this eroding of distances. The implosion of space – the distant becomes near via the Internet – is a characteristic of the electronic age. The television turns the World into a 'global village' through the images it sends from all over the world into one's living room (see McLuhan 1968). At the same time – and here lies a decisive difference to the television – the Internet challenges individuals to act: an aspect reinforced by Web 2.0. The Internet and the computer/smartphone/tablet/smartwatch, etc., demand participate. The Internet requires a new cultural practice of participation. The Internet is experienced as the alien, which demands user action through a content/form linkage: The Internet requires active users and specific communication practices (e.g., blogging, writing wiki articles, etc.) to unfold its decentralized communication potentials adequately. These specific communication practices can be understood as a new way of dealing with the World or new forms of production. From this perspective, the Internet represents the medially foreign, which conveys foreign information from other cultures through digital media. The Internet, or rather the (cultural) practices it requires in order to open up the World of the digital, challenges one's own world. These theoretical considerations can be empirically established based on the DIVSI Internet milieus.

On behalf of the German Institute for Trust and Security on the Internet (Deutsches Institut für Vertrauen und Sicherheit im Internet; DIVSI), the Sinus Institute Heidelberg (Sinus-Institut Heidelberg) has analyzed the online behavior of the German population with recourse to the Sinus Milieu Model. The Milieu Study on Trust and Security on the Internet was first published in 2012 and updated in 2013. Latest data is available for 2016. As part of the milieu study, different ways of accessing the Internet were based on a population-representative typology. Against the background of a milieu-analytical approach, the DIVSI study measures people's perceptions of safety with regard to the Internet. The study results allow access to the socio-economic implications or the 'structuralist constructivist' dimension of the Internet: How do people construct the Internet as a space of experience against the background of the socio-economic structures in which they are embedded? (Fig. 1.1)

The DIVSI Internet milieus from 2016 show that the Internet is (still) perceived in part as a foreign medium. Thus, the milieu of the so-called 'Internet-distant insecure' can be used to show how the Internet is perceived as a medium that enters one's 'own life' as something foreign and thereby evokes an effect of the uncanny. Such a perception of the Internet as something alien can be seen paradigmatically in statements on the perception of security by actors who have been identified as representatives of the milieu of the 'Internet-distant insecure':

> And then, with these social networks like Facebook, you always hear: 'Oh, you can write to me, I'm on Facebook.' I said: 'No, I won't do that.' Because you hear and read so much about what can happen there, you can be spied on Facebook. (female, 67 years) (quoted from DIVSI 2016, p. 72).

The Internet represents the uncanny threat, an unforeseeable threat, and has been brought by hearsay. The Internet is experienced as a new dangerous cultural space. The authors of the DIVSI study state that the actors of this milieu, which is the

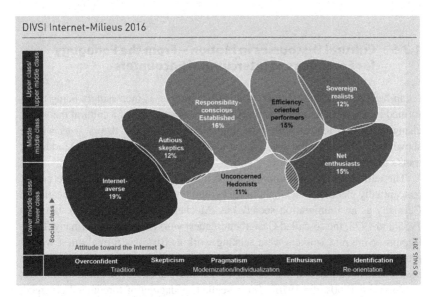

**Fig. 1.1**  DIVSI Milieu study on trust and security on the Internet, DIVSI 2016, p. 31

largest milieu with 19%, find "the digital world alienating all in all" (DIVSI 2016, p. 71). However, a comparison of the milieu of the Internet-distant insecure between 2011 and 2016 reveals significant shifts. In 2012, for example, this milieu had a share of 27%. The Internet is increasingly losing its status as a stranger and is inscribing itself into society's media structure. As one of the significant changes in the digital lifeworld from 2012 to 2016, the authors of the study highlight that "the Internet [...] has arrived in the middle of society" (DIVSI 2016, p. 31):

> A pragmatic, unagitated approach to the Internet has become normal for large sections of society, and uncertainty and skepticism play a smaller role in possible digital attitudes than they did in 2012. The Internet's value has also increased among offline users, although their security concerns have also grown (DIVSI 2016, p. 31).

The Internet is increasingly being integrated. However, this integration process has socio-economic implications. Another finding of the 2016 DIVSI study is that the "digital society [...] is drifting apart" (DIVSI 2016, p. 31). Thus, "[f]or example, in 2012 there were still significantly more people not only in a simple but also in an upscale social situation among those who were reserved and distant from the Internet" (ibid.). In 2016, on the other hand, "those distant from the Internet are now almost exclusively people with low incomes and little formal education" (ibid.).

## 1.3.5 Cultural Discourses in Motion – From the Pedagogy for Foreigners to Intercultural Encounters

It can be stated that the Internet, as a media stranger, is increasingly being transformed into an own. In this transformation process, models of a cultural theory are changing, which increasingly have implications for media theory – as will be shown below based on the differentiation-theoretically oriented cultural models of trans-culture and hyper-culture.

The cultural theoretical approaches of trans-culture and hyper-culture can be conceptually elaborated against the background of the discourses on education in Germany as an immigration society (see Mecheril et al. 2010, p. 56). Between 1955 and 1973, the so-called Gastarbeiter/'guest worker' came to Germany. These foreign or migrant workers were seeking work as part of a formal guest worker program (Gastarbeiterprogramm). In course of the migration process, education issues were not addressed in a 'decade of discursive silence' (Mecheril et al. 2010, p. 56) in the 1960s, the 1970s. The so-called 'pedagogy of foreigners' emerged:

"The integrative pedagogy for foreigners, which was oriented towards the assimilation of young migrants, assumed that the immigrant children and adolescents had considerable socio-educational deficits, which primarily manifested themselves in linguistic deficits that had to be compensated for" (Kiesel n.d., p. 4). The deficits result from the difference between the foreign, the 'foreigner children' (see Mecheril et al. 2010, p. 56), and the own. Symbolically, the foreign is marked by linguistic competence, which actualizes the cultural meaning of language that Herder pointed out in his cultural philosophy. In the 1990s – the "decade of difference discourse" (Mecheril et al. 2010, p. 56) – difference-theoretical approaches influenced the theory on culture.

The 'starting point' of this influence is the so-called intercultural pedagogy. The central impetus for establishing a difference discourse was the "insight that 'our foreign fellow citizens' are not a temporary phenomenon in society" (Mecheril et al. 2010, p. 56). "Concepts that consider intercultural education and upbringing not as additional special treatment for 'foreigners' children' but as, for example, a self-evident school task is being developed" (Mecheril et al. 2010, p. 56 f.). In the course of this process, intercultural pedagogy was conceived as a strategy of cultural encounter. This encounter space forms the 'inter-' of intercultural encounters. In this context, a totality-oriented understanding of culture is often actualized. The individual functions as the representative of a culture:

> Culture is a universal orientation system typical for a society, organization, and group [...]. This orientation system is formed from specific symbols and is passed on in the respective society, etc. It influences the perception, thinking, values, and actions of all its members. It influences all its members' perception, thinking, values, and actions and thus defines their membership in society. Culture as a system of orientation structures a field of action specific to the individuals who feel they belong to society and thus creates the conditions for developing independent forms of coping with the environment (Thomas 1993, p. 380).

This totality-oriented concept of culture and the logic of representation (the individual as the representant of a culture) lead to the conceptualization of an individual encounter between two individuals attributed to different cultures as an intercultural encounter. One effect of such a conceptualization of cultural encounter is to essentialize culture: A particular culture is defined by specific characteristics ascribed to the respective actors, from which the actors cannot escape. The individual is culturalized through an intercultural lens – **"culturalization** means describing social relations, population groups, affiliations and social ties with the help of ethnic categories and highlighting cultural differences" (Kabis 2002, p. 2, e.i.o.). In culturalization in the context of intercultural pedagogy, foreign ethnicization occurs – for example, when a child with 'African roots' is asked to bring a 'typical food' from his home country for a school festival. According to this logic, the foreign 'African' culture is represented by a culture-specific food: African culture is encountered on the school festival's food buffet. This logic of intercultural pedagogy assigns the child a specific nation/continent as well as to a specific (food) culture and thereby constructs him or her as a representative of the foreign – "If children and adolescents are seen primarily as representatives of a particular culture, this presupposes the assumption that people are trapped in their culture, that they are, so to speak, in a 'cultural dungeon' from which they could not free themselves" (Kabis 2002, p. 3). In intercultural pedagogy, children with a so-called 'migrant background' are discursively thematized as strangers and placed in difference to the 'own' culture. At the same time, intercultural pedagogy opens up a space for the discourse of difference. The 'inter-' forms the formulation of an understanding of cultural encounters beyond dichotomous juxtaposition and creates a liminal understanding of the culturally foreign. The foreign and the own mix in a threshold state (see Turner 1998) – an aspect that Bhabha has elaborated in cultural theory from a postcolonial perspective.

The beginning of this discourse of difference in the early 1980s occurs at a time of increasing computerization: Prensky (2001) defines anyone born from 1980 onwards as a digital native, as the increasing digitization of society has socialized them. At the same time, 1980 is the year in which Reagan becomes President of the United States and begins to impose neoliberal policies that are still effective today. The establishment of a discourse of difference thus occurs at a time of fundamental social transformations, which are also inscribed in the way we deal with the Internet and construct performatively digital cultures. Simultaneously, approaches to cultural theory oriented towards the theory of difference, such as hyper-culture, deal with digitally-based practices of bringing forth.

## 1.3.6   Trans-culture – Cultural Patchwork in the Digital Age

Welsch formulated the concept of trans-culture in the 1990s – at the time of the first commercialization of the Internet. Welsch continues to pursue this model and elaborates it over the next years. The model of trans-culture is based on the consideration that "[c]ivilizations [...] are conceivably strongly interconnected and interwoven" (Welsch 2010, p. 3): "Most among us are determined in their cultural formation by *several* cultural origins and connections. We are cultural hybrids. The cultural identity of today's individuals is a patchwork identity" (Welsch 2010, p. 5). This approach abolishes the intercultural logic of representation that defines an individual as a representative of a culture. Instead, Welsch posits that "[t]he way of life of an economist, a scientist, or a journalist [...] is no longer simply German or French, but – if anything – European or global in character" (Welsch 2010, p. 3.). In doing so, Welsch distinguishes trans-culture from a traditional under-standing of culture, which he locates with Herder:

> This spherical model includes an internal homogeneity imperative and an external demarcation imperative. Internally, culture should shape the life of a people as a whole and individually and make every action and every object an unmistakable com-ponent of precisely *this* culture; foreign elements are minimized in this conception. And in the external reference, strict delimitation applies: every culture, as the culture of *a* people, should be specifically distinguished and distanced from the cultures of *other* peoples (Welsch 2010, p. 2).

With the overcoming of the so-called 'sphere model', the foreigner as a delimiting instance disappears. Implicitly, the root model of cultural descent is replaced by an 'inter- 'and an overlapping of cultures. This overlapping and interlacing of cultures and the accompanying dissolution of the foreign become possible through indi-viduals' transcultural imprinting (see Welsch 2010, p. 5). Through such trans-cultural imprinting, dichotomous demarcations between the own and the foreign erode: through an 'inner trans-culturality' (Welsch 2010, p. 6), individuals do not define something different as something foreign. Instead, the encounter of the for-eign in the other refers to one's transculturality.[8]

> An individual into whose identity a whole series of cultural patterns has found its way possesses greater chances of connection with regard to the multitude of cultural prac-tices and manifestations found in his social environment than if his own identity were determined by only one pattern (Welsch 2010, p. 6).

Following this logic, transculturality facilitates cultural encounters. Welsch points out that "the more elements an individual's cultural identity is composed of" (Welsch 2010, p. 6), "the more likely it is [...] that there is an intersection with the identity of other individuals" (ibid.). These intersections enable "individuals, for all their other differences, to engage in exchange and communication to a far greater extent than before; they can discover existing commonalities and develop

---

[8] In the sense of a provisional distinction, it can be stated that transculture as a term encom-passes the conceptual dimension, while transculturality denotes the concrete state of a cul-tural patchwork.

new ones" (ibid.).[9] Welsch locates transculturality beyond the "blinkers" of a "thinking of delimitation" (Welsch 2010, p. 7) – "trans-culturality liberates one to make one's own choices beyond socially predetermined schemata" (ibid.). In this ethical perspective on trans-culture, the relationship between individual/culture is redefined: Whereas for Herder, the individual still functions as a representative of culture, according to Welsch, the individual is a bearer of multiple cultural elements that are updated according to the situation (this results in a certain proximity of Welsch's approach to role theory). Trans-culture can thus be understood as a manifestation of a difference-theoretical concept of culture. Instead of a primary collectivity (the individual represents a culture), a multi-collectivity (the individual constitutes a cultural patchwork) occurs. The trans-culture approach shows a postmodern opening of cultural totality claims. A redefinition of culture also accompanies this postmodern opening. It could be said that a difference-theoretical concept of culture replaces a normative concept of culture.

The normative concept of culture is defined by an evaluative juxtaposition of cultural/aesthetic phenomena/objects preserved through the formation of tradition and form a cultural heritage. High culture is defined as a canon of aesthetic works and 'great' artists. Everyday culture and popular culture are often excluded or devalued in the process. In the course of this, as in the so-called evolutionary understanding of culture, a hierarchization of cultures can occur. The evolutionary understanding of culture represented a form of the normative concept of culture and was formulated in the nineteenth century. Within the framework of such an evolutionary understanding of cultural development, the topos of a civilized European culture functions as a normative reference point. Other cultures are supposed to develop in the sense of progress towards European culture:

> Civilization is a plant much often propagated than developed. As regards the lower races, this accords with the results of European intercourse with savage tribes during the last three or four centuries; so far as these tribes have survived the process, they have assimilated more or less of European culture and risen towards the European level as in Polynesia, South America (Taylor 1871, S. 48).

In the sense of colonial discourses of legitimation, the term culture rationalizes hierarchy and dependency relationships. Such a normative concept of culture can be contrasted with a difference-theoretical concept of culture in which several cultural patterns are inscribed in the individual. Such a difference-theoretical concept

---

[9] From this perspective, transculture can be interpreted as a postmodern understanding of culture, since various transcultural constructs exist alongside, with, and within each other in the sense of postmodern petits récits, in mutual tolerant recognition.

of culture is based on the implosion of time and space. This medial implosion of time and space characterizes the electronic age: The telephone enables possible synchronous communication that is no longer bound to space. Due to the implosion of distances caused by telecommunication, cultures are increasingly 'detached from the ground.' From a media-theoretical perspective, the difference-theoretical concept of culture is based on the electronic age infrastructure.

With the enthronement of the individual as the bearer of a multi-collectivity, there is, at the same time, a secularization of cultural truth claims. In the cultural patchwork, cultures and their value systems are confronted with truth claims of other cultures and thus relativized. Herein lies the ethical dimension of trans-culture. Cultural truth claims' relativization represents a rational thinking and an epistemologically critical approach, which is the fundament of transcultural knowledge. Within transcultural thinking, the modern enlightened individual functions subtextually as the epistemological point of reference in the sense of transcendental epistemology, the individual becomes aware of his or her cultural inscriptions, can relate to the truth claims of the respective cultural contexts in a relativizing way, and reflexively moderates his or her own transcultural identity or the various cultural inscriptions. The epistemological model of a rational subject is actualized, which was conceptualized in German Idealism and shaped the epistemological deep structures of (European) modernity and the understanding of the bourgeois individual (see Kergel 2011a, b, 2013). From this perspective, a transcultural stance requires a Eurocentric account of the epistemic subject. Like Lyotard's model of postmodern epistemology, Welsch's cultural opening reproduces a culturally prefigured, modern concept of bourgeois epistemology and consequently remains culturally bound. This subtextual enthronement of a European-influenced epistemological approach can be problematized as the implicit Eurocentrism of trans-culture.

Another central criticism of Welsch's model of trans-culture is its neglect of a power-analytical perspective. With reference to Zizek (1999), it can be stated that Welsch's approach refrains from a power-analytical differentiation of sociopolitical aspects. Thus, Zizek (1999) refers to

> two entirely different socio-political levels, on the one hand, the cosmopolitan academic from the upper class or the upper-middle class, who always crosses borders with the right visa without any problems to pursue his (financial or academic) business in different countries, enjoying the differences; on the other hand, the (im)migrant worker who has been driven out of his homeland by poverty or (ethnic, religious) violence, and for whom the vaunted 'hybridity' denotes the, traumatic experience of never being able to properly settle down and legalize his status (Zizek 1999, p. 155).

The postcolonial discussion focuses on this desideratum and places the power-analytical dimension of cultural identity construction at the center of cultural-theoretical reflections. This postcolonial perspectivization of a difference-theoretical understanding of culture leads the dissolution of the foreign towards a subversive diversity that unfolds in the course of the communicative potentials of the digital age. Despite these two points of criticism – enthronement of a Eurocentric subject of knowledge and neglect of the power-analytical perspective – the trans-culture model provides an understanding of culture beyond a dichotomous per-spective on cultural encounters. These cultural encounters are thereby shaped by the electronic age's medial structure or by an implosion of space and time. Although Welsch states that transculturality always characterizes cultural identities, he re-curs to globalization, which he identifies as an intensified driver of transcultural interconnections: The starting point is the thesis that globalization is replacing ho-mogeneous cultural identities with transcultural encounters. In this context, Hepp (2015) points out that with "globalization [...] new constructions of ethnicity" (Hepp 2015, p. 300) emerge, "which no longer necessarily refer to the cultural identity of supposedly homogeneous nations" (ibid.). Instead, artifacts that mark cultural identity become citations or traces of origin employing "a global postmod-ernism borne through media products" (ibid.).[10] Welsch's claim is to respond to these media changes in cultural theory with the model of trans-cultur (see Welsch 2010, p. 1 f.). The cultural changes also arise from an implosion of culture. This implosion is characterized by the fact that "more and more [...] the same articles (however exotic they may once have been) become available everywhere" (Welsch 2010, p. 3). Here Welsch updates an analytic topos developed by McLuhan in his engagement with the electronic age: Technological change, distant places move into one's place. Welsch transfers this logic to cultural exchange processes: Changes in "worldwide systems of transport and communication, as well as global capitalism" (Welsch 2010, p. 4) remove a territorial binding of culture. The implo-sion of space, paradigmatically represented by the television in the electronic age, is extended by the Internet and radicalized at the latest with the appearance of Web 2.0 and its decentralized structures, its poly-directional, and its polyphonic poten-tial. Consumers of media content from all over the world enter into a dialogue with other actors from across the globe via digital media and constitute what Han (2005) calls a hyper-culture. Hyper-culture can be defined as an understanding of culture

---

[10] At the same time, nationality aspirations can be observed in the European states, which manifest in the success of right-wing populist parties and thus help the spherical understand-ing of culture to gain new popularity.

according to which cultures emerge in the synchronous action of individuals. This action is an interplay of several cultural patterns. This dissolution of boundaries leads to hyper-cultural diversity.

### 1.3.7   Hyper-culture – De-localization and De-removal of Culture

With reference to the technological progress of the digital age Han (2005) reformulated the patchwork approach of difference-theoretical cultural models such as trans-culture: "The process of globalization, accelerated by new technologies, removes cultural space. The resulting proximity creates an abundance, a fund of cultural life practices and expressions" (Han 2005, p. 17). Han emphasizes that "cultures implode, they become hyper-culture" (ibid.). As in the model of trans-culture, "cultural contents" "crowd" one another in a juxtaposition, they "interpenetrate" and "overlap" one another (ibid.): diverse cultural contents are available to all places at all times.

Analogous to the trans-culture model, the thesis could be put forward that the implosion of spave is also accompanied by cultural de-totalization: the claim to totality that can be inherent in cultural life practices is structurally negated – hyper-culture secularizes culture. In this process of secularization, culture is also detached from the soil: "Culture becomes genuinely cultural, indeed *hyper*-cultural, by being de-naturalized, by being freed from both 'blood' and 'soil,' i.e. biological or *terran* codes" (Han 2005, p. 17, ed.). Instead of assuming a territorial attachment of culture to a place, Han speaks of the 'eros of interconnectedness' regarding the model of hyper-culture (see Han 2005, p. 19). "It is not the sense of trans-, inter- or multi, but hyper- that more accurately reflects the spatiality of contemporary culture" (Han 2005, p. 17). This spatiality is defined by an implosion of space that transcends the interconnectedness of the electronic age:

> Globalization does not mean that the village is networked with the here. Rather, globalization produces a global here by removing and de-locating the village. Neither interculturality nor multiculturality, nor transculturality can mark this global here. The hyper-cultural tourist travels the hyperspace of events that opens up to cultural sightseeing. Thus, he experiences culture as a cul- tour (Han 2005, p. 47, ed.).

The individual becomes a cultural traveler. Through the implosion of space, the culturally foreign as a concept erodes. From this perspective, hyper-culturality "presupposes certain historical, socio-cultural, technical or media processes" (Han 2005, p. 60).

Han elaborates on the spatial understanding of the networking concept underlying the hyper-culture model. He refers to Nelson's understanding of hypertext and Deleuze and Guattari's rhizome model – aspects that develop a central significance in the context of the theoretical examination of Web 2.0.

Hypertext is "not limited to the level of the digital text" (Han 2005, p. 15) but is defined by a 'struc-tangle.' Referring to the term tangle, which can be translated as tangle or knot (see Han 2005, p. 15), Han sees the struc-tangle of hypertext as "a structured tangle" (Han 2005, p. 15). According to Han, this structured tangle characterizes cultural reality and constitutes a hyper-structure of culture. This hyper-structure of culture is defined by the fact that culture "increasingly [loses] that structure which resembles that of a conventional text or book" (Han 2005, p. 16). Against the backdrop of globalization and technological change, which leads to an implosion of cultures or removal of cultural space, cultures can no longer develop a claim to totality. In the sense of postmodern discourses, culture can no longer function as a metanarrative: "No history, no theology, no teleology makes it appear as a meaningful, homogeneous unity. The boundaries or enclosures to which the semblance of cultural authenticity or originality is attached dissolve" (Han 2005, p. 16).

Han transfers his analytical description of cultural reality into an ethical sphere: in the sense of a postmodern logic that identifies potentials for freedom by playing with petits récits, (small narratives), Han sees spaces of freedom in the decentered structure of the hypertext and the rhizome: while "[l]inear and hierarchical structures or closed, unchanging identities [...] represent results of constraint, hypertext 'promises' a "freedom from constraint" (Han 2005, p. 15). Against the background of these considerations, hyper-culture does not represent a monoculture but can be understood as a principle of cultural networking (Han 2005, p. 22). In this networking, cultural practices, ways of life, and forms of expression emerge that are constituted by a composition of different cultural elements. To illustrate this process of constitution leading to hyper-culture, Han chooses the metaphor of 'fusion food.' Fusion food, or fusion cuisine, "is a mixed cuisine" (Han 2005, p. 23). This fusion cuisine mixes something new with recourse to the "hyper-cultural fund of spices, ingredients, and forms of preparation" (Han 2005, p. 23). This does not negate the "diversity of food cultures" (ibid.). Instead, new forms are "created" (ibid.). Fusion cuisine evokes, "a multiplicity that would not be possible if local cuisine were kept pure" (ibid.).

Han develops a hyper-cultural concept of identity that actualizes features of the postmodern subject: In a hyper-cultural "constellation of being" (Han 2005, p. 54), the potential for freedom of self-location in the play of narratives or cultural expressions. That leads to the absence of a "gravitation that unified parts into a binding wholeness. Being disperses into a hyperspace, of possibilities and events that, instead of gravitating, merely buzz within it, as it were" (ibid.). According to Han, this potential for freedom effects a "decline in the horizon" that can be experienced nihilistically "as an aching void, as a narrative crisis" (ibid.). Simultaneously, this decline in the horizon of hyper-cultural space "also opens up a new practice of freedom" (Han 2005, p. 54). In this process, one's identity is 'pieced together' from a "hyper-cultural fund of ways of life and practices" (Han 2005, p. 55). "Thus, patchwork-like entities and identities emerge. Their polychromy points to a new practice of freedom that owes itself to the hyper-cultural defactification of the life-world" (Han 2005, p. 55). According to Han, in the fusion cuisine of hyper-culture, cultural practices mix with each other – they are linked continuously anew. This dynamic is the freedom of hyper-culture. This freedom represents the ephemeral dimension of hyper-culture, which is also a significant feature of digital interaction or digital cultures. Just as hypertext enables diverse, decentralized networking of content, hyper-culture represents a decentralized, non-hierarchical practice of cultural patchwork in performative movement.

To illustrate this non-hierarchical, decentralized practice, Han draws on the rhizome model developed by Deleuze and Guattari. Han interprets Deleuze and Guattari's rhizome model as a hyper-cultural linking processes which is realized on a medial level through hypertext. According to Han, the rhizome refers to "the uncentered multiplicity that is not subject to any overarching order" (Han 2005, p. 32). From a cultural theory perspective, the rhizome can be understood as a counter-design to a taxonomic model of cultural roots that ties culture to a place. The family tree is substituted by the rhizomatic network of hyper-culture.

It is precisely the hyper-cultural approach of the de-familiarization or de-localization of culture that proves to be connectable to the rhizome model: "Hyper-culture as de-internalized, de-rooted, de-localized culture behaves rhizomatically in many ways" (Han 2005, p. 33). Like hyper-culture, "which is not a culture of interiority or memory" (Han 2005, p. 34), the rhizome "has no memory" (ibid.). Instead of preserving memory, what occurs is an ephemeral constitution of hyper-cultural identities. From this perspective, the hyper-cultural praise of the de-located culture also necessitates the problematization of "a cultural diversity-oriented" approach "towards protecting species, which could only be achieved through artificial fencing. Unfruitful would be the museum or ethnological plurality" (Han 2005, p. 22). Instead, "the vitality of a cultural exchange process [...] includes the spread, but also the disappearance of certain forms of life" (Han 2005, p. 22). The 'hyper-cultural identity' (see Han 2005, p. 54) is defined by the fact that "[a]ny monochrome self [...] is replaced by a multicolored self, a *colored self*" (Han 2005, p. 55, ed.). Unlike trans-culture, for example, there is no longer an alien that can be appropriated hermeneutically:

The hyper-cultural tourist is not, after all, a *hermeneuticist*. Hyper-culture also differs from multiculture in that it has little *memory of* origins, ancestry, ethnicities, or places. And for all its dynamism, hyper-culture is based on a dense juxtaposition of different ideas, signs, symbols, images, and sounds. It is a kind of cultural hypertext. Trans-culturality lacks precisely this dimension of the hyper. It is not the vastness of the trans, but the closeness of the Spatio-temporal juxtaposition that characterizes contemporary culture (Han 2005, p. 59, ed.).

The foreign is a construction, part of an "often destructive separation between the own and the foreign" or the "old and the new" (Han 2005, p. 62), and has no place in hyper-culture: in terms of a dichotomous tension, however, Han positions hyper-culture against a 'fundamentalism of place': hyper-culture presupposes a "readiness for difference, for the new" (Han 2005, p. 62). This attitude enables the decentralized performative dynamics that characterize hyper-culture. Accordingly, Han emphasizes that "[t]he hyper-culture is without center, without God, without place" (Han 2005, p. 18). According to Han, hyper-culture's secular decentralized claim means that its approach "will continue to provoke resistance" (Han 2005, p. 18): "For not a few, after all, it leads to the trauma of loss. Retheologization, remythization, and renationalization of culture are common phrases against hyper-culturalization of the world" (ibid.). Han holds that "de-localization will continue to confront a fundamentalism of place" (ibid.). Despite this dichotomous tension between the old and the new, de-location and renationalization, Han claims that the hyper-culture approach does not need a power-analytic perspectivization. Thus, postcolonial cultural criticism is abandoned, which is characterized by the concepts of hybridity and third space developed by Bhabha are characterized. As Han notes, "Hybridity, however, by its very conceptual history, is too closely tied to the racist and colonialist complex of power, domination, oppression, and resistance, to the geometry of center and margins or top and bottom. Thus, it fails to capture the very playfulness" (Han 2005, p. 30). With the abandonment of the power-analytical perspective, hierarchy and dependency relations are lost from an analytical view, which, according to postcolonial cultural analyses, shape precisely the constitution of cultural self/world relations. On the other hand, with the model of hyper-culture, Han outlines a space of possibility for a largely power-free, postmodern cultural practice in the digital age. Web 2.0, or the poly-directional and polyphonic ephemeral 'fluidity of the digital,' can evoke a decentralized, rhizomatic hyper-culture.

With hyper-culture, Han draws a picture of a cultural practice of the Internet or Web 2.0 that enables a non-hierarchical, decentralized 'accretion of spaces' (Han 2005, p. 30). These spaces would be 'accessible not in terms of power economy but in terms of aesthetics.' (ibid.). Such an aestheticist drawing deliberately neglects a power-analytic engagement with (digitally-based) cultural practices. This approach neglects the power struggles, the establishment, and subversive undermining of hegemonic structures, which are also inscribed in digital cultures.

In summary, it can be stated that the concept of culture is a concept in becoming. This becoming is also defined by historical and medial aspects. When social constellations change, what is understood by culture also changes. Against the background of Herder's understanding of culture, cultures never appear in the singular: Culture has a logic of demarcation, since culture is always different from other cultures. This logic of demarcation can be analytically worked out heuristically with Simmel's model of the foreign. Wherever culture is spoken of, the logic of demarcation is subtextually at work, and what actualizes other cultures' presence. This relationship between other cultures is discussed in approaches of inter-, trans- and hyper-culture. In the course of the discourse, it can be stated that the model of the culturally foreign is abandoned in favor of a cultural plurality that inscribes itself in the individual. Individuals thus constitute trans- or hyper-cultural patterns of identity: individuals create their own culture. In their cultural-theoretical argumentations, Welsch and Han emphasize that globalization de-locates cultures. More or less explicitly, they also refer to the media implications of cultural change. The implosion of cultural spaces is accomplished in the electronic age – for example, through the television. In the digital age, this implosion is extended by the participatory, action- and production-oriented structure of digital cultures. It is precisely the media-induced cultural plurality that implicitly takes up patterns of argumentation that have been defined within the framework of philosophical postmodernism: Cultural truth claims are substituted in favor of a juxtaposition and interrelation of different cultural practices. This juxtaposition leads to a multiplicity or diversity of cultural practices in the sense of trans- and hyper-culture. This diversity of trans- and hyper-cultural practices is accompanied by a relativization of cultural truth claims. The abandonment of truth claims in favor of an interplay of plurality is epistemologically reworked in the context of postmodern epistemology with the model of metanarratives or grand narratives and petits récits. Following, postmodernism is reconstructed as an epistemology developed in the context of the electronic age. With the decentralized polyphonic possibilities as well as poly-directional possibilities of Web 2.0, postmodern communication potentials unfold, which are part of digital cultures and which have a subversive diversity.

# Postmodern Cyberspace 2

## 2.1 From Modernity to Postmodernity

The discussion of postmodernism can be identified in a wide variety of fields and ranges from literature, architecture, and epistemology to cultural theory. At the same time, the term postmodern has a semantic range that leads Eco (2003) to understand it as a 'passe-partout' (Eco 2003, p. 77). In the following, the term postmodern is understood less as a designation as an era than as an epistemological approach. From this perspective, postmodernism represents a critical-subversive approach to the claims to totality and normative tendencies that unfold in modernity. 'Postmodernism' consequently results from the relational structure of modernism/postmodernism. In the sense of a differential interplay of demarcation, both terms define reciprocal – "Although there are significant differences between the Postmodernist theories can all focus on the criticisms of modernism" (Yaakoby 2012, p. 10).

From the perspective of media theory, the Internet has a communication structure that enables postmodern social practices. The formation of postmodern social practices can be understood as metonymic manifestations of a "postmodernist cultural change" (Angermüller 2013, p. 246). The postmodern cultural change has been prepared by the media infrastructure of the electronic age. The thesis can be formulated on this basis,

- that telecommunications, which shaped the electronic age, required a postmodern perspective on the world or construction of reality and that this
- postmodern construction of reality is unfolding via the Internet and in digital cultures.

© Springer Fachmedien Wiesbaden GmbH, part of Springer Nature 2023
D. Kergel, *Digital Cultures*, https://doi.org/10.1007/978-3-658-35250-9_2

This thesis will be unfolded below. To define the notion postmodernism, a conceptual understanding of modernity will first be outlined.

## 2.1.1   Modernity – Society in Transformation

From a discourse-analytical perspective and in the sense of a schematization of social developments, modernity can be understood as the state of a society that – caused by socio-economic and socio-cultural shifts – has abandoned a 'traditional' self-understanding. The fundamental epistemic bases[1] of such a self-understanding discourse has its modern roots in the "Querelle des Anciens et des Modernes" (1687). In the poem "Le siècle de Louis le Grand," by Perrault, the exemplary function of antiquity was relativized in favor of a modern state. The poem formulated a dichotomy between the old and the modern.[2] As with the concept of postmodernity, it should be noted that modernity has field-specific characteristics and is therefore defined differently by different scientific fields. From the perspective of the state's history, modernity is often identified with the Renaissance and the French Revolution (on state theories in historical change see Zippelius 2003, as a central work identifying modernity with the modern age see Tönnies 2010). On the other hand, in German literary historiography, it is above all the literature of the late nineteenth century that is modern literature or literature of the modern age. The plurality of literary styles alone – Impressionism, Neo-Romanticism, Symbolism, Expressionism, New Objectivity – some of which existed alongside one another, some of which followed one another in quick succession, reflects the dynamics of a society in flux, struggling to understand itself (see Mix 2010, on the sociological dimension of literature, see Bourdieu 1999).

From a historicizing, sociological perspective, modernity can be understood as a self-understanding discourse of bourgeois society. The following discussion of modernity focuses on the discursive self-determination of Western European bourgeois societies. The discursive self-identification as modern, which is at the same time a redefinition of a society as a secular, industrialized society, is strategically

---

[1] According to Foucault (1974), an episteme describes how interpretations of the world are prefigured and which categorizations are updated in the acts of interpretation. An episteme is not conceived as a category. From this perspective, an episteme thus includes which categories lie outside the conceivable of a historical situation (such as a possible death of God in the twelfth century).

[2] This dichotomy is still influential today and, among other things, has an impact even in social state modernization theories (see Lessenich 2012).

demarcated from a traditional self-understanding. In the course of this demarcation, the social condition is defined as new or modern. Social changes cause such a process of demarcation. In the modernization of European society, fundamental transformation processes are taking place. These transformation processes produce a need for the renewal of social self-interpretation. These self-interpretations manifest themselves in the thematization of modernity as a discourse of self-understanding for society as a whole. This discourse was influenced not only by technical progress and industrialization but also

- through a profound change in the structure of employment, which led to a strengthening of the tertiary sector and thus also placed increased demands "on the expectations of the overall structure of qualifications" (Tenroth 2000, p. 189),
- by employment crises and unemployment caused by the cyclical dependence of capitalist economies: "While unemployment was virtually unknown until 1914 (but an uncertain age perspective and exploitation are quite common in the life course), employment crises since 1918/1929 have become the rule rather than the exception" (Tenroth 2000, p. 189),
- The increase in population, which is mainly based on the extension of life expectancy and the decline in infant mortality (see Tenroth 2000, p. 193, this also leads to a cross-field discursive thematization of urbanization).
- By mostly problematic housing situations: "The average apartment around 1900 was a one-room apartment (plus kitchen) in large cities and working-class families" (Tenroth 2000, p. 192).

In addition to these factors, there are also media upheavals that shape the living environment and the discourse landscape. In several phases, the media landscape is fundamentally renewed by the assertion of different mass media: the newspaper is followed by radio, and in the 1930s, albeit still sporadically, by television. All these factors fundamentally restructure the world in which we live, and with it the socializing processes of individuals: "With the beginning of the 20th century, the social and political, educational and biographical conditions of growing up overlap in a peculiar, contradictory form" (Tenroth 2000, p. 203).

In summary, it can be said that in the course of the discourse on modernity, the model of progress already formulated in the modern era and the Enlightenment increasingly established as a metonymy of modern society. In the course of this discourse, modern society dichotomously demarcating itself from the old and its traditions. The self-understanding discourses of modernity/modern society can also be understood as approaches to (re-)establishing coherent world views. They are, therefore, a discursive, meaningful response to social acceleration and transformation. These transformations force a new social self-understanding since the traditional matrices of meaning and interpretation patterns no longer appear valid: Modernity is also an effect of an ever more complex living environment and appears in the process of discursive self-assurance already in crisis. Paradigmatically, the crisis-like nature of modernity is shown in the so-called language crisis: the language crisis as an identity crisis was already problematized by the Viennese poet Hugo von Hofmannsthal at the turn of the nineteenth and twentieth centuries. One could say that a modern language crisis began symbolically in 1900 when Hofmannsthal published the fictitious letter of a fictitious Lord Chandos. Chandos cannot give his life a coherent expression, so that "words fall apart in his mouth like musty mushrooms" (Hofmannsthal 1951, p. 14). Language and cohesive experience of life or reality are intertwined. It appears as an indication of a social change that Chandos' letter's appearance falls into a time in which many life-reform initiatives were developed.[3]

Lukács explicitly address a connection between artistic language and the social state of society in his 1916 essay "The Theory of the Novel" ("Die Theorie des Romans"): "The novel is the epic poem of an age for which the extensive totality of life is no longer evident, for which the life immanence of meaning has become a problem, and which nevertheless has an attitude toward totality" (Lukács 1971, p. 47). The direct connection between language and work of art is also exemplified in modernist discussions of the novel, where the disintegration of language into different codes is discussed: "The 'crisis of the novel' – this expression is increasingly encountered – would also exist today if there were no novel at all. It is a crisis of our reality in general" (von Doderer 1959/1972, p. 85). The novel's crisis is a crisis that was also caused by the fact that a totalizing narrative, which discursively assigns a totalizing meaning to social transformation processes, no longer appears to be generatable.

In the literary field, experiences of crisis in modernity are articulated, which Lyotard perspectives epistemologically in his 1979 book "The Postmodern Condition." This writing, which Lyotard wrote on behalf of Quebec/the government's University Council, focuses on post-industrial society. According to Lyotard, modernity is characterized by unifying narratives/total social interpretations of the world. These narratives/world interpretations view the social transformation processes teleologically as a process of progress. Lyotard gives the concept of postmodernism an epistemological foundation.

---

[3] In 1900, for example, the "Wandervogel" were founded, in 1903 the first German nude bathing resort was opened in the Bay of Lübeck, and in 1900 Carl August Heynen founded the health food store in Wuppertal-Barmen.

## 2.1.2 Metanarrations and Postmodern Subversion

From Lyotard's perspective, modernity is discursively identified as "the age of scientific, political and economic systems ideologically secured by ideals, by 'grand narratives,' by 'meta-narratives'" (Niedermair 1992, p. 87). Metanarrations such as the Enlightenment or Marxism, create a coherent understanding of society. They can be understood as an ordering strategy by placing the past, present, and future in a meaningful relationship. History is not experienced in a nihilistic way but is given a teleological concept through metanarrations – which is why these metanarrations often have a "salvation-historic[al] character" (Baum 2010, p. 87). From this perspective, metanarratives are part of bourgeois society's self-understanding discourses: Modern metanarratives reflect the "culture of modernity" (Preyer, n.d., p. 10) – they are "evidence of a [...] typical value system, the components of which have been characterized by the intrinsic values 'universalism' and 'individualism,' 'activism' and 'rationalism.' They should be institutionalized in a 'rational society'" (Preyer n.d., p. 10).

Metanarrations constitute a monolithic offer of social coherence that guarantees claims to truth: "True knowledge [...] is always indirect knowledge; it is composed of reported statements that are incorporated into the metanarrative of a subject that guarantees their legitimacy" (Lyotard 1983, p. 35). Lyotard's "philosophic[al] postmodernism" (Niedermair 1992, p. 88) contrasts these metanarrations with the model of petits récits. As with the language crisis or the novel's crisis, the starting point is the loss of the legitimacy of coherence. Thus, Niedermair (1992) emphasizes that the "philosophical postmodernism [...] accentuates a counterpoint to modernity" (Niedermair 1992, p. 88). At the same time, "[t]he manifold figures of the legitimation of human action by ideals [...] can no longer be accepted without question" (ibid.). According to Lyotard, postmodernism is characterized by the adoption of metanarrations, and, from this perspective, it does not so much occupy an epochal status. Instead, postmodernity can be understood as a state of mind (see Yeh 2013, p. 38). "Postmodern consciousness is for Lyotard consciousness of a rupture" (Georg-Lauer 1988, p. 191). This consciousness of a rupture is defined by a tension between metanarrations and petits récits: Lyotard sets the model of the petits récits against the metanarrations. Instead of totalizing metanarrations, a discourse landscape of petits récits is proposed, which enables a plurality/diversity. Petits récits function as a critical contrast to the totalizing claim of metanarrations. Through their diversity, they have a subversive potential beyond metanarrations, since they "draw attention to the mechanisms of exclusion [...] that take effect wherever there is 'narration' with a legitimizing intention" (Baum 2010, p. 88). This subversive epistemological perspective is marked with the prefix 'post-.' The 'post-' in the constellation of modernity/postmodernity marks the break with the totalizing claims to coherence and totality of metanarratives. The epistemological perspective, which is metonymically signaled by the prefix 'post-,' can be placed in a line of tradition with Adorno's concept of negative dialectics – there is always something that cannot be grasped conceptually or narratively through the metanarratives. A postmodern attitude is characterized by the epistemological awareness that social actors are not adequately represented in the discourses. In addition, through petits récits, social actors can develop a voice in the context of social self-understanding discourses. The hegemonic claims of metanarrations are subversively undermined by the diversity of petits récits. Deleuze and Guattari (1992) sum up this interplay between metanarratives and petits récits in their book "Thousand Plateaus" ("Tausend Plateaus"). Deleuze and Guattari point to the interplay between minority and majority: "Majority implies a constant of expression or content that is something like a yardstick by which it is evaluated" (Deleuze and Guattari 1992, p. 147). The majority is defined by a "state of power and domination" (ibid.). From this perspective, the majority presupposes "the standard" (ibid.). The majority is not an empirical quantity but a discursively produced hegemonic image:

Let us suppose that the constant or standard would be the male-white-adult male who lives in cities and speaks some standard European heterosexual language [...] It is evident that the 'man' has the majority, even if he is less numerous than mosquitoes, children, women, blacks, farmers, homosexuals, etc. (Deleuze and Guattari 1992, p. 147).

The minority, on the other hand, is characterized by the fact that it is discursively assigned a "determination other than the constant" (Deleuze and Guattari 1992, p. 147). Again, this is "no matter what kind and how large it was, that is, it was regarded as a sub-system or external system" (ibid.). Metanarrations function in this conceptual approach as the majoritarian, "as a homogeneous and constant system, the minorities as sub-systems and the minoritarian as a possible, creative and created one" (ibid.). This conceptual logic captures a process of postmodern genesis of narratives: Deleuze and Guattari exemplify this postmodern processual interplay of sociolects: "Black Americans do not confront *black* with English; they make a *black-English* with the American, which is their own language. Mity languages as such do not exist: they exist only in relation to the high-level language" (Deleuze and Guattari, 1992 p. 146, ed.).

The minority is given a voice through short narratives. Accordingly, the postcolonial theorist Bhabha (2011) sees in the postmodern concept of petits récits also the empowerment of marginalized groups. Thus, according to Bhabha, the "broader meaning of the postmodern situation [...] is rooted in the realization" (Bhabha

2011, p. 6) that the "epistemological 'boundaries'" of ethnocentric ideas "also represent the articulatory boundaries of several other dissonant, even dissident stories and voices" (ibid.):

> As the demography of the new internationalism consists of the history of post-colonial migration, the narratives of the cultural and political diaspora, the tremendous social displacements of peasant and indigenous communities, the poetics of exile, the gloomy prose of refugees for political and economic reasons (Bhabha 2011, p. 6 f.).

The dissolution of the metanarratives is also reflected in the fact that "concepts such as homogeneous national cultures [...] are currently fundamentally redefined" (Bhabha 2011, p. 7). Petits récits metonymically refer to the foreign, which cannot or will not be captured by a metanarration. Petits récits disturb a totalizing order and are therefore extraordinary. The symbolic order of metanarrations and their claims to hegemony are undermined by petits récits, making them subversive and evoking a postmodern disorder.

## 2.1.3 Order of Modernity and Postmodern Disorder

The power-critical perspectivation of modern social self-understanding discourses turns postmodernism into a cultural theory or cultural philosophy project. Postmodernism has the effect of redefining the concept of culture; it 'decenters culture' (see Buckow 2008, p. 128). Bhabha (2011) stresses that this power-critical dimension is formally given in the prefix 'post-,'. The 'post-' does not represent overcoming a state, but rather a shaping of the critical potentials, a further thinking or 'going beyond':

If the jargon of our time – postmodernism, postcolonialism, postfeminism – has any meaning at all, it is not in the popular use of the post as an expression of a sequence – *postfeminism*; or a polarity – *anti*-modernism. These terms, which emphatically refer to the 'beyond,' embody its restless and revisionist energy only when they transform the present into an expanded and eccentric place of experience and appropriation of power. If, for example, interest in postmodernism is limited to celebrating the fragmentation of the 'great histories' of post-Enlightenment rationalism, it remains a deeply limited endeavor despite its intellectual stimulation (Bhabha 2011, p. 6, ed.).

From this power-analytical perspective, the interplay between modernity and postmodernity can be conceptualized as a play of order and disorder. Reese-Schäfer (1989) proposes to understand 'postmodernism' as "the avant-garde experimentation in the age of its origin" (Reese-Schäfer 1989, p. 46), "where it goes against conventions, against consensus, and against taste" (ibid.). On the other hand, modern refers to the "attempt to deliver unity, realism, to establish orders" (ibid.). Angermüller (2013) sees in postmodernism the potential to build, develop, and perpetuate modernity (see Angermüller 2013, p. 259). At the same time, postmodernism makes it possible to react to the crisis of the loss of a unifying narrative, to which Hofmannsthal's Chandos letter metonymically refers. Postmodernism is the awareness that no reconciliation between different language games is expected (see Angermüller 2013, p. 259). The *subversive implications of petits récits or the critical-reflexive rupture with orders marked by the prefix 'post-' can be understood as an epistemological foundation of the cultural-analytical approach to subversive diversity.* Thus, Buckow (2008) points out that "[t]*he dislocation and decentration of culture* [...] *opens up new opportunities for diversified cultural discourses"* (Buckow 2008, p. 138, ed.): Petits récits undermine totalizing claims to hegemony, and in each case, the unattributable is not excluded but recognized in its singularity. In these considerations lies an ethical dimension of postmodern epistemology.

## 2.1.4  Postmodern Ethics as Reflective Tolerance

Each narrative has its premises, discourse logics, and value systems within which rational action unfolds. The ethical consequence lies in the imperative to recognize the codes/language games of other narratives as different but of equal value (see Lyotard 1983, 64ff.). This approach results in an epistemological attitude in which narratives do not clash narratively but enter into a power-free discourse in a recognizing and inclusive manner. In their work on postmodern education – "Postmodern Education: Politics, Culture, and Social Criticism" – Giroux and Aronowitz (1991) describe such a tolerant attitude as an educational goal:

> A tolerant self-reflexivity, a positive appreciation of the self-narration – thereby knowing about its relativity – can be seen as one of the basic educational goals of a postmodern education program. The values that constitute postmodern education are those of empowerment in the most profound meaning of the term (Giroux and Aronowitz 1991, p. 22 u. p. 109).

The demand for tolerant self-reflection as an educational goal of postmodern education shows significant characteristics of postmodern ethics: According to Baumann (1995), postmodern ethics is characterized by the renunciation of (culturally) founded normative rules of action, since these cannot be justified in terms of postmodern relativism. It cannot be determined how and according to which standards the effects of ethical action are evaluated. The consequence is,

> to live without such guarantees and with the awareness that they will never exist either – that perfect society, as well as an ideal human being, is not a feasible prospect and that attempts to prove the contrary lead to greater cruelty than to more humanity – and certainly to less morality (Baumann 1995, p. 23).

The premise – which is not mentioned in this quotation – is to acknowledge the conditionality of other life models or petits récits and to suspend their claim to universal validity – and the claim to the universal validity of one's own (cultural) identity. The normative implications of culturally legitimate ethical rules and forms of moral action are substituted by an 'indifference' (see Rancière 2008) of petits récits. To avoid a struggle for the establishment of hegemonic culture, a reciprocal knowledge of the 'indifference' of all narratives appears as a premise of postmodern ethics. From this perspective, postmodernism can also be understood as an attitude of a tolerant encounter. In this encounter, critical questioning of truth claims must also be made (see Kergel 2017b). If such an epistemological-critical perspective is not adopted, the questioning that characterizes a postmodern attitude to knowledge erodes.

As an attitude of knowledge, the term 'postmodernism,' unlike in other fields such as architecture or painting (see Lyotard in Reese-Schäfer 1989, pp. 110 f.), does not refer to the end of modernism as an epoch. "What, then, is the postmodern? [...] It is undoubtedly a part of the modern" (Lyotard 1983, p. 79). Thus, as a thinker of postmodern epistemology, Lyotard emphasizes in an interview that the term "'postmodern' for me does not mean the end of modernity, but a different relationship to modernity" (Lyotard in Reese-Schäfer 1989, p. 112). From this perspective, the interplay between modernity and postmodernity results from a tension between hegemonic homogenization through metanarrations and subversive heterogenization or *diversity* through petits récits. A postmodern epistemological attitude is defined by a tolerant attitude toward other narratives and questioning of truth claims. In his analysis, Lyotard refers to IT technology's postmodern potential to realize such a tolerant, dialogical communication process. At the same time, Lyotard also emphasizes the potential of 'computerization' for domination and subjugation. Thus,

> the computerization of society [...] the "dream" instrument for controlling and regulating the market system, extended to include knowledge itself and governed exclusively by the performativity principle. But it could also aid groups discussing meta perspectives by supplying them with the information they usually lack for making

knowledgeable decisions. The line to follow for computerization to take the second of these paths is, in principle, quite simple: give the public free access to the memory and data banks (Lyotard 1983, S. 67).

Computerization and the digitalization can promote the postmodern dialogue that is the communication structure of postmodern ethics. Simultaneously, computerization can take the form of technology of domination of the control society. Lyotard is in line with the Open Source movement and the Creative Commons approach in his demand to release information.

## 2.1.5   The Postmodern Subject, Dialogue as Postmodern Knowledge and Eurocentrism in Postmodern Epistemology

The logic of a dialogical postmodern tolerance presupposes the epistemological reflection of truth claims: Only the bracketing of truth claims enables an indifference toward other narratives. Such an epistemologically critical attitude was developed in modern times by Descartes with the formulation of strategic doubt. The postmodern subject updates the critical strategy of doubting by questioning truth claims. In the course of this questioning, the autoreferentiality of knowledge is simultaneously established: Through strategic doubt, the subject constitutes itself as an autoreferential instance of knowledge:

> It's thinking. It alone cannot be separated from me: I am, I exist, that's certain. [...]
> But how long have I been? Well, as long as I think. [...] So I am just a thinking thing
> (res cogitans), i.e., spirit (mens), soul (animus), intellect (intellectus), reason (ratio)
> (Descartes 1972, p. 20).

Knowledge is only possible through doubt (see Kergel 2011b). Descartes establishes an epistemological approach that, via Kant and German idealism, influences the modern epistemological self-understanding of bourgeois society and provides the epistemological foundation of the postmodern subject. This foundation of the subject is epistemologically defined by doubt. An exploratory relationship to the world characterizes the subject. This exploratory relationship or behavior is manifested pre-reflexively in an explorative behavior and cognitively in a strategic questioning of truth claims. Descartes elaborates this cognitive process in terms of an epistemological foundation of the subject. In a second step, the subject constructs reality on the epistemological basis of doubt. Thus, a constructivist reading of Descartes is based on the thesis that doubt constructs an understanding of the world by the subject. From this perspective, the subject is the agent of world construction. Doubt does not destroy knowledge but enables a doubting/critical and open-minded relationship with the world. Kemmerling (1996) points out that Descartes' axiomatic version of auto-referentiality is to be understood as being *in itself* and *for itself*: "The ego – the individual human mind, the individual human soul – is a substance, and that means with Descartes: a thing that exists in such a way that it does not need any other thing (except God) to exist" (Kemmerling 1996, p. 103). Sesnik (2014) sees a revolutionary potential in a subject, which constitutes itself through a critical attitude: The social world and thus also established hierarchies and relations of dependence are strategically questioned:

> The subject constitutes itself in the rebellion against reality. It is per se a revolutionary subject, and as such, it creates free space for a new world in which fate is no longer imposed. Now it can finally give itself its law, that is, become autonomous. The world surrounding it and the world that it is itself has dissolved it; it enters the space between no longer and not yet (Sesnik 2014, p. 37).

The subject stands in a critical and examining relationship to truth claims. The discursive legitimation of hierarchies and relations of dependence is challenged. From this perspective, the doubting subject is also a subject that rebels against truth claims. This perspective has a long tradition. For example, Schelling points out in the context of German idealism that reflexive self- and world-knowledge becomes possible through critical autoreferentiality. According to Schelling, the possibility of world-knowledge moves man into the proximity of God: "The consequence of things from God is a self-revelation of God. But God can only reveal himself in what is similar to him, in free beings acting out of themselves; for whose existence there is no reason but God, but who are as God is" (Schelling 1809/1964, p. 57). The critical and examining attitude is ambivalent. For where the subject functions as the auto-referential agent of knowledge, it puts itself into a recognizing or critically examining relationship to reality or God. In Schelling's epistemological diction, the critical-reflexive attitude toward the world and oneself represents a sin, since reflection violates the truth claims of metanarrations (such as 'God') and can thus be evaluated as a heretical epistemological practice: "Thus, the beginning of sin is that man passes from the actual being into non-being, from truth into a lie, from light into darkness, to become a self-creating ground, and to rule over all things with the power of the centri that he has in him" (Schelling 1964, p.108). In the course of these subject-philosophical reflections, the subject becomes a critical instance of knowledge that constructs the world and thus itself through doubt. Concerning Descartes, who founded doubt as a form of autoreferential epistemology for modern philosophy, the stocks of knowledge and claims to truth become precarious. Based on Descartes' doubt and in the course of subject philosophy, the postmodern subject's epistemological cornerstones as a doubting subject are pre-configured: *The postmodern subject constitutes itself in the critical examination of the world in a social context.* *This epistemological attitude of the postmodern subject is defined by Zima (2007) as ephemeral – and thus shows proximity to the characteristics of the digital*: "The individual subject is neither something sovereign-fundamental nor subject, but rather a changing, semantic-narrative and dialogical entity that lives from the confrontation with the other, the stranger" (Zima 2007, p. 88, ed.). The subject becomes 'fluid' by continually constructing new stocks of knowledge, critically examining them, and in the course of this falsifying, modifying, or expanding them. In other words: the postmodern subject does not 'rest' on predetermined stocks of knowledge. In the constant critical examination of these knowledge stocks, epistemological agility of the subject is constituted performatively.[4] Such an understanding of autoreferential knowledge is not to be understood as the self-assurance of a monologizing subject. Instead, doubt unfolds epistemologically in a dialogue or polylogue in which perspectives on the world come together. Subjects critically discuss interpretations of the world and themselves. Dialogue is the form of communication of doubting world construction and requires a mutual understanding (see Kergel 2017b). Only the other's position, the adoption of perspectives, the critical and examining bracketing of one's own truth claims and the truth claims of the interlocutors enable a dialogue. *From this perspective, dialogue can be defined as the form of communication of the postmodern attitude.*

The postmodern construction of a doubting self/world relationship enables the subject to locate himself beyond the normative-subjectivating compulsion of meta-narrations. This freedom is problematized by Angermüller (2013) when he states that in 'postmodernism,' the subject finds it difficult "to build up a stable sediment of historical or social experience: it does not develop a sense of a historical before or after or a socio-structural up and down" (Angermüller 2013, p. 250). The hyper-cultural approach formulated by Han sees in this form of cultural throwing into the pluralism of postmodern petits récits the possibility of a 'timeless' culture that is not bound to traditions and has no cultural heritage:

> The hypertextually written world consists, as it were, of countless windows. But none of the windows opens an absolute horizon. But this lack of horizon-like anchoring of

---

[4]This epistemological agility of the subject in constant motion is formally manifested in Hegel's "Phenomenology of Spirit." In this major work of Hegel's, metaphors and verbs are often used, which semantically denote 'fluidity.'

enabling a new way of walking, a new perspective. In *windowing*, one glides from one
window to the next, from one possibility to another (Han 2005, p. 54 f., e.i.o.).

While Angermüller remarks that "the subject is as it were directed at the mercy of
the aesthetic-sensual impressions in postmodernism" (Angermüller 2013, p.
250), Han understands this as a potential for aesthetic-sensual freedom. Significantly,
Han uses terms such as hypertext and windowing to update an inventory of terms
that is significant for the Internet's infrastructure. Besides, Han formulates the hyperculture approach when the poly-directional and polyphonic potentials of the
so-called Web 2.0 are increasingly becoming part of the Internet culture. The communication potentials of Web 2.0 make it possible to realize the media dimension
of a postmodern dialogue. Beyond this media-theoretical perspective, it should be
emphasized that postmodern tolerance is based on a doubting or critically examining (self-)objectification of truth claims. This doubting attitude enables a tolerant
self-reflexivity that characterizes postmodern encounters. Such an understanding
of postmodern encounters has far-reaching consequences: Postmodern reflection is
a manifestation of a doubting, self-objectifying, and relativizing insight, which,
with Descartes' approach, stands at the beginning of modern epistemology. From
this perspective, postmodernism is a culmination of modern epistemological strategies. Through postmodernism, modernity enlightens itself by undermining the hegemonic claims of metanarrations: "Postmodernism's critique of modernity is es-
sentially not directed at modernity in general, but solely at the spirit of modernity
as the spirit of the great concepts of unity" (Barz 2011, p. 3). From this perspective,
postmodernism represents a *Eurocentric* epistemological attitude since postmodern doubt cannot transcend the context of Western epistemology. Thus, even in a
cultural-theoretical turn of postmodern epistemological logic, Eurocentric implications remain: The opening for other cultures is based on an epistemologically skep-
tical, rational attitude of reflection, which stems from a specific cultural sphere –
the epistemological discourses of European modernity (see Kergel 2011a). Almost
imperceptibly, the cultural opening through a postmodern attitude becomes an enthronement of Western epistemology and its normative implications. From this
perspective, postmodernism, contrary to Habermas's (1993) fears, nevertheless
holds on to the 'project of modernity' – a postmodern paradox that can be endured,
but probably not resolved, from the perspective of cultural encounters.

  A central possibility to formulate a postmodern attitude of knowledge is postmodern protest. A protest is a form of articulation (see Laclau and Mouffe 2012).
The diversity of petits récits can be formulated through the act of refusal and resistance to normative-hegemonic access. From this perspective, the postmodern attitude of knowledge is an attitude of protest. The protest as a form of questioning

delimitation remains in the symbolic order, which it can only reflexively transgress employing negativity. As an act of negation, however, protest cannot leave the symbolic order, since negativity dissolves without a frame of reference:

> Negative thus conveys a restlessness that disturbs the slumbers of the given; that undermines any reified plenitude, presence, power, or position, be it symbolic or political. It is *affirmative* in its engendering (creative, although sometimes in a destructive mode), yet *negating* (critical, transgressive, subversive, etc.) vis-à-vis the positive. As such the motif of negativity itself refuses definition in terms of any simple binary opposition (positive/negative) (Coole 2000, S. 74, e.i.o).

Thinking of negativity, which can be understood as rebellious thinking because of its subversive attitude toward the symbolic order, becomes a productive part of system-immanent transformation processes (see Kergel 2013). The postmodern protest that negates truth claims represents the action form of postmodern disorder vis-à-vis modern order. It is precisely the understanding of negative or rebellious thinking as a driver of (social) dynamics – "the rhythms of negativity that choreograph the dialectic" (Coole 2000, p. 52) – that can be understood as stabilizing the system because of its system immanence. The protest remains within the symbolic order points to the potential for improvement inherent in the system and can transform symbolic orders inherent in the system.

## 2.1.6    Irony as a Postmodern Protest – From Schlegel, via Spontaneous Slogans to Twitter

The first publicly documented message between computers dated from 1969 when the social protest or civil rights movements of the 1960s and 1970s were increasingly articulated. The 'unrest at the universities' (Gilcher-Holtey 2008, p. 25) – for example, in Berlin and Berkeley – was a forum for anti-authoritarian protest. Student groups such as the 'Students for a Democratic Society' were committed to organizing universities as forums of resistance:

> Seeing the youth as a new carrier and the universities as 'potential bases and agencies of a movement for social change,' they advocated that the universities should attempt to act against indifference in society, to point out the political-social and economic causes of private difficulties and public grievances, and to develop alternatives and possibilities for change (Gilcher-Holtey 2008, p. 19 f.).

The universities provided intellectual freedom, which Derrida later summarized with the model of an 'unconditional university' (Derrida 2015). The academic freedom, the possibility of thinking differently, also accelerated the development of the internet:

> The culture of individual freedom that flourished at universities during the 1960s and 1970s used computer networks for its purposes – and mostly sought technological innovation for the sheer fun of discovery. Universities themselves played an important role by supporting community networks (Castells 2005, p. 34).

The Internet also emerged in the context of the protest movements of the 1960s and 1970s. The anti-authoritarian dimension of these protests made it possible to formulate the perspectives of a *'culture of individual freedom'*. Following Schneider (2004), the anti-authoritarian protest movements can be understood as a postmodern protest. The fact that the postmodern attitude is also a politically actively participating attitude or a critical one is also elaborated by Baumann (1995): "Postmodernism is not the end of politics, just as it is not an end of history. On the contrary, everything that is attractive about the postmodern promise calls for more politics, for more political commitment, for more political effectiveness of individual and communal action" (Baumann 1995, p. 339). As an epistemological approach, postmodernism has a critical position toward the hegemonic claims of metanarrations. The petits récits of postmodern skepticism are brought into a position in protest against these claims to hegemony. The public rejection of the hegemonic claims of metanarrations can be understood as a postmodern protest, which can manifest itself in various forms: These range from aesthetic self-reflection (see Foucault see 1989) to subversive anarchist protests (e.g., the so-called pudding assassination of the Commune 1) and forms of artistic subversion (e.g., Dadaism). The protest can be defined as a public manifestation of resistance, rejection, refusal, and as a form of negation: Hessel (2011) closes his pamphlet "Outrage" with the words "To create something new means to resist. To resist means to create something new" (Hessel 2011, p. 10). The refusal is an expression of a desired alternative and can positively affect the self/world relationship: "I wish all of you, every single one of you, a reason for indignation. That is precious. When you are outraged about something [...] you become active, strong, and committed" (ibid.). *As a form of resistance protest are a central feature of the tension between the hegemonic claims of metanarrations and the subversive diversity of petits récits that enter the discourse through protest.*

A characteristic of postmodern protest is the ironicization of hegemonic claims to power and their assertion of interpretative sovereignty. As a form of protest irony creates an epistemologically space for petits récits. Noetzel (2008) speaks of the "susceptibility of postmodern politics to irony" (Noetzel 2008, p. 46). At least on the epistemological level of romantic irony, a critical epistemological attitude toward epistemological claims to hegemony is formulated: "A philosophical writing cannot end with mere rh(etoric). The conclusion must be ironic (annihilating or ironizing)" (Schlegel 1980, p. 98). While the "unironic refers to the objectivity of a general truth" (Noetzel 2008, p. 42), the distancing form of irony allows questioning truth claims.[5] In epistemological terms, irony was effectively formulated in the context of early Romanticism by Schlegel and others. In Romantic irony, the writer reflects on himself and his literary work. In this way, a differentiation, a bracketing of one's position, is performed. At the same time, a critical perspective on knowledge is realized, which is typical for postmodern epistemology: An epistemological characteristic of Romantic irony consists in a relativization of ultimate knowledge. This relativization leads to epistemological agility/an infinite dynamic of knowledge which can be interpreted as a fundamental attitude of an ephemeral consciousness: "Irony is a clear awareness of the agility(,) of the infinitely full chaos" (quoted from Strohschneider-Kohrs 1967, p. 77). Chaos – this is how this passage could be interpreted – can be represented as the present or the world dissolved into diverging, individual elements. The subject gains distance employing Romantic irony. Schlegel speaks of a "joke where one destroys oneself again and again" (Schlegel 1980, p. 105). Modern thinking of totality is undermined by the attitude of romantic irony – "The ironist puts everything into it and takes everything back, because only in this way can he understand himself as the execution of a fundamental reference to the comprehensive whole, which always transcends him and his achievements" (Bubner and Hörisch 1987, p. 92). The ironic self-objectification, which does not take itself seriously, enables a play with truth claims. With reference to Hegel the form of ironic destruction of meaning can be defined as a comedy. In his "Lectures on Aesthetics" ("Vorlesungen über Ästhetik"; 1835–1838), Hegel elaborates the epistemological meaning of irony and comedy. According to Hegel, in comedy, subjectivity makes itself "the master of all circumstances and purposes" (Hegel 1976, p. 547). "[S]o it is in comedy [...] subjectivity, which in its infinite certainty, has the upper hand. [...] in comedy, in the laughter of the individuals who dissolve everything through themselves and in themselves, we see the victory of their subjectivity, which nevertheless stands securely within itself" (Hegel 1976, p. 552). Comedy represents a relativization of all values and thus enables the solution of normative attributions of meaning:

---

[5] Here Bourdieu (2015) sees the subversive power of sociology when he states: "Sociology resembles comedy, which reveals the hidden mechanisms of authority" (Bourdieu 2015, p. 86).

> The general ground for comedy is, therefore, a world in which man as a subject has made himself the complete master of all that which is otherwise regarded as the essential content of his knowledge and accomplishment: a world whose purposes are therefore destroyed by its lack of essence (Hegel 1976, p. 552).

Hegel critically problematizes the subversive dimension of irony: the relativity evoked by comedy makes the constitution of meaning impossible. Meaning can be destroyed but not deconstructed through humor: in acts of ironic demarcation from God, this point of reference and, therefore, the basis of demarcation always remains a constituent part of the comedy. From this perspective, critical irony, like protest per se, remains an immanent part of the criticized system. Due to the bracketing of truth claims, irony can be understood as a specific form of postmodern resistance: By ironically breaking with truth claims, by strategically questioning any ultimate justification, a postmodern dynamic performed.

Ironic, postmodern protest was formulated in the wake of the 1968 movement in Germany by the left-wing alternative milieu of the Spontis through so-called Sponti slogans. The left-wing alternative Sponti movement can be recognized as an undogmatic left position, whose decentralized forms of the organization find their digital continuation in the cyberspace of the Internet:

> The term 'Sponti' was derived from the leftist understanding of spontaneist/ spontaneistic and dealt with the appropriate relationship between organization and spontaneity. According to the Spontis, forms of fighting and life were to be directly connected. Theoretical work and experience should be related to each other. Against dogmas and party discipline, a voluntaristic and subjectivistic understanding of politics was put into action. "Sponti" was initially a swearword, but by the mid-1970s, it had become a positive description for a left-wing party that was primarily involved in the milieu and was distrustful of institutions, organized groups, and planned actions. Instead, spontaneity, self-organization, practice, and autonomy became the new magic words (Reichardt 2014, p. 115).

The Sponti movement led to alternative culture in cities such as "Bremen, Göttingen, Freiburg, Marburg, Heidelberg, Hamburg, and Berlin" (Reichardt 2014, p. 117). This alternative culture led to the "establishment of an autonomous infrastructure of its own, ranging from pubs, alternative newspapers, and craft service cooperations to the municipal and children's shop movement" (ibid.). The undogmatic position of the left-wing Sponti milieu manifests metonymically in so-called spontaneous slogans. These sayings are often a variation of idiom or proverbs modified in the sense of ironic refraction. These spontaneous sayings were chanted at demonstrations and were found as graffiti on walls or public toilets.[6] Sayings such as "Better juso than ouzo" or "Better nudism than FDP" are characterized by a provocative hedonism. This provocative hedonism undermines the claim to seriousness and, at the same time, by equating hedonistic life practices with political discourses, understands these life practices themselves as political. The political discourse is subversively infiltrated and enriched by an ironic life-affirming perspective.

Such a form of ironic protest finds its continuation in the virtual world of cyberspace. One example is the artist duo "Yes Man," consisting of Jacques Servin and Igor Vamos. The first known action of the Yes Man was located in the material-physical world and consisted in the exchange of the voice recordings of play dolls:

> Unnoticed by numerous sellers, in 1993, the hackers managed to exchange the voice recordings of 80,000 Barbie dolls with those of the mannequin 'GI Joe.' As a result, Barbie dolls were sold that demanded to go to war, while the soldier dolls 'GI Joe' suggested to go shopping. An education in *gender studies* that the manufacturer Mattel would certainly have gladly done without (Bardeau and Danet 2012, p. 43, ed.).

The Yes Man extended this form of subversion to the virtual world. For example, Servin was employed by Maxis. This company developed the simulation game Sims. In this PC game, which sold millions of copies, players can simulate an everyday world. Servin had to "leave the company because he had taken arbitrary initiatives by inserting a scene into the game in which two men in swimming trunks kiss" (Bardeau and Danet 2012, p. 43). This form of ironic subversion was also extended to the Internet when the Yes Man secured the domain "gatt.org" (General Agreement on Tariffs and Trade) and published a fake website of the World Trade Organization (WTO) on the Internet: "They succeeded in creating the impression that on the Yes-Men websites they operated, the ones they were actually 'talking' to were those to whom they created proximity by semantic means" (Schönberger 2006, para. 15). Thus, e-mails and invitations to congresses were sent to the alleged website, which the artists accepted. At the congresses, the artists then appeared with excessive demands. The Yes Man shows how 'spontaneous resistance' is gradually shifting to the Internet:

> The Yes Men stand for net activism that uses the Internet's technical conditions to develop new or expanded (recombinant) forms of protest. At the same time, they

---

[6] See in terms of a collection of Sponti sayings Domzalski 2006 or the website http://staff-www.uni-marburg.de/~naeser/sponti.htm, last accessed: September 2, 2017.

conceive of the Internet as a political sphere of action with a specific understanding of politics. A prerequisite for the boost in significance that such actions have received in recent years is, however, that there is a critical mass of those "symbol analysts" who have the corresponding cultural capital and can carry out such forms of ambitious protest. The IT sector's growth and corresponding services form the social prerequisites for this (Schönberger 2006, para. 34).

The ironic subversion is inherent in a culture of communication that undermines hegemonic powers of interpretation in the sense of postmodern knowledge. The decentralized structure of the Internet seems to accommodate irony as a form of postmodern protest and knowledge. The Internet and the ephemeral communication culture make it possible to form a suitable media structure for irony. Hoinkis (1997) refers to the relevance of.

Irony as an attitude toward oneself as a person as well as toward the social environment, both for the author and the recipient, is not or only to a limited extent possible based on communication that implies a high degree of physical presence, involvement in the communicative process, normative orientation and thus little scope for free and individual arrangements of redundancies (Hoinkis 1997, p. 45).

Irony requires movement or a space that cannot be directly countered repressively. This space of freedom is made possible by the Internet. An example of the Internet-based ironic communication culture can be seen in Twitter reaction to the so-called alternative facts.

The formulation "alternative facts" was made by Kellyanne Conway, an advisor to US President Donald Trump, in January 2017. She used it to justify Sean Spicer's statements, the White House Press Secretary, during a political talk show ("Meet the Press"). Sean Spicer had claimed that Trump's inauguration had been attended by a larger audience than that of Barack Obama, and any other inauguration. This assertion could not stand up to the empirical facts – including aerial photographs of Obama's and Trump's inaugurations. Asked about this, Conway replied that Spicer would have given alternative facts to the one-sided reporting against Trump. At this point, an attempt was made to enforce the hegemonic claim to interpretation: Dialogical negotiation processes about perceptions of reality were avoided by formulating alternative facts. From the position of governmental power, an attempt has been made to exercise interpretive power. Instead of a dialogue about positions, a post-factual form of politics was enforced. The enforcement of positions defines such a post-factual policy through emotionalization, whereby a rationally based discourse is suspended. In this specific case, the form of post-factual argumentation did not remain unanswered. Using the hashtag '#alternativefacts,' the term 'alternative facts' was exposed as a rhetoric of post-factual politics by means of an ironic decontextualization and exposed to ironic ridicule in the sense of postmodern protests (Figs. 2.1 and 2.2).

As a Web 2.0-based communication medium, Twitter made it possible to give a voice to actors who otherwise – unlike newspapers, politicians, etc. – have no voice. Through the hashtag, the ironic protest became supra-individual, and through the de-location of the Internet, the protest could not be fixed. This medial form of ironic protest via Twitter took place on a semantic level through an articulatory strategy that stands in the tradition of spontaneous slogans: A decontextualization ironizes the contrast between hegemonic claims to truth and repressive seriousness with seemingly profane counterparts. It is possible to reconstruct a genealogical structure that ranges from Romantic irony to left-wing alternative Sponti movement to an ironic protest via Twitter.

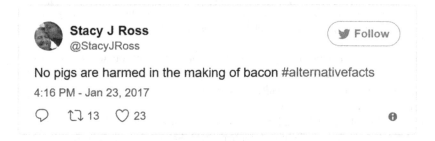

**Fig. 2.1** Ironic Twitter reaction to Conway's strategy of legitimizing postfacto politics. (Source: http://www.huffingtonpost.com/entry/twitter-alternative-facts_us_58860e9be4 b070d8 cad3b1ec, last accessed October 10, 2017)

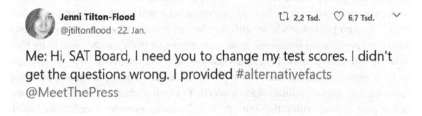

**Fig. 2.2** Ironic Twitter reaction to Conway's strategy of legitimizing postfacto politics. (Source: https://twitter.com/i/moments/823217993429549056?lang=de, last accessed on October 10, 2017)

## 2.1.7 The Emergence of Subversive Diversity – Postmodern Protest in the Third Space

Postmodern protest can also be used for cultural theoretical analyses. The abandonment of a coherent understanding of culture leads to approaches that conceptualize an interrelatedness and an interlocking of culture. Formally speaking, this conceptualization process is reflected in the use of prefixes such as 'inter-' (intercultural), 'trans-' (trans-culture) and 'hyper-.' (hyper-culture). The image of a uniform culture is undermined in the sense of the concept of petits récits: The conception of a cultural interlocking abandons the model of a large coherent leading culture. If power-critical aspects are excluded from postmodern approaches like trans-culture and hyperculture, or at least are not at the center of the analysis, postcolonial cultural analyses focus on the interaction between the 'majoritarian' and the 'minoritarian'. Bhabha, for example, states about the "cultures of a postcolonial *countermodernity*" (Bhabha 2011, p. 9, ed.) that they are "in an adjoining, discontinuous or oppositional relationship to modernity" (ibid.). In this way, they can "resist their oppressive, assimilating techniques; but they also use the cultural hybridity." (Bhabha 2011, p. 9) With postcolonial cultural criticism, an essentialist concept of culture is substituted by a difference-theoretical concept of culture: An essentialist concept of culture assumes that a sum of definable characteristics determines cultures. These include language, customs, and traditions. Culture is often equated with nation and functions as a metanarration. The interpretations of a normative and essentialist concept of culture are undermined by a power-analytical, difference-theoretical understanding of culture. For example, instead of analyzing the encounter between colonizing imperial power and colonized culture as an intercultural encounter, cultural identities are constituted in the encounter. This encounter constitutes a liminal space – in Bhabha's words: a third space. Laclau and Mouffe (2012) point out that only in the encounter, "the colonizer [...] is discursively constructed as the non-colonized" (Laclau and Mouffe 2012, p. 168). The identity assignments colonized/non-colonized is the effect of an encounter in the context of which cultural identities only emerge through discursive demarcation dynamics between the colonized and the non-colonized: "In this sense, the border becomes the place from where something *begins its essence*; this happens in a movement that resembles the unsteady, ambivalent character of the connection with that which lies beyond" (Bhabha 2011, p. 7).

Bhabha conceives this moment of a power-pervaded cultural encounter as a third space (see Bhabha 2011, p. 326ff.): Cultural dichotomies dissolve and open up a liminal free space play of differences of demarcation and identification. Thus, "cultures are never to be understood as uniform, and cultural borders are thus not to be taken for granted, but rather as fields for negotiating difference" (Kerner 2012, p. 127). To describe this process, Bhabha uses the metaphor of the staircase:

> The staircase as a threshold space between identity determinations becomes a process of symbolic interaction, a connecting structure that constructs the difference between above and below, black and white. The staircase's back and forth, the movement and transition in time that allows it, prevent identities from becoming fixed at its upper or lower end to original polarities. This inter-spatial transition between fixed identifications opens up the possibility of cultural hybridity in which there is a place for difference without an adopted or decreed hierarchy (Bhabha 2011, p. 5).

Such an understanding of culture as a third space "avoids polarities and contingencies" (see Castro Varela and Dhawan 2015, p. 269). The third space opens up narrative spaces for minorities that are otherwise defined by metanarrations. The third space thus enables a "history of post-colonial migration" (Bhabha 2011, p. 6) as well as a discursive space for petits récits or minority "narratives of the cultural and political diaspora" (Bhabha 2011, p. 6). The postcolonial approach of a third space reveals the subversive dimension of a power-analytical postmodern understanding of culture: petits récits are not fixed, but are constituted in confrontation with the hegemonic claims of the majority and thus subvert these claims to power – "the ambivalence of the third space gives rise to the ability to act and the potential for subversion" (Kerner 2012, p. 129). *In the third space, a subversive diversity consisting of petits récits is thus generated. This subversive diversity is implicitly based on the premise that the petits récits are also heard or that the subjects can articulate themselves in petits récits.* This unnamed premise updates media-theoretical aspects and considerations of the media infrastructure of postmodern communication processes.

## 2.1.8 The Struggle for Publicity – From the Monologue of Television to Web 2.0-Based Dialogue

Protest requires publicity. The creation of a public sphere can be defined as a central challenge to the articulation of protest. This also raises the question of the appropriate media for postmodern protest: "Media create public spheres; this is an aspect of their social character. Without media, there would be no public sphere at all" (Münker 2009, p. 29). Along with the change in the media, there is also a change in the public sphere and, in the forms of protest. Based on the change of protest forms, the media shifts that lead from the electronic age to the digital age can be worked out paradigmatically.

Thanks to the poly-directional and poly-phonic potentials of Web 2.0, protest and petits récits can be formulated with lower thresholds than was the case, for example, in the 1960s and 1970s, when television was the leading social medium. As a mass medium, television had a normative interpellative function and was thus able to produce hegemonic images of metanarrations. Thus, Stalder (2016) points out how homogeneous the spectrum of participants in the television discussions of the 1950s and 1960s was:

> Mostly white, heteronomous men who held important institutional positions in the West's centers spoke to each other. As a rule, these were highly specialized actors from culture, business, science, and politics. Above all, they were legitimated to appear in public, articulate their opinions, and see these opinions as relevant and discussed by others. They conducted the important debates of their time. Other actors and their divergent positions, which of course always existed, were, with few exceptions, either categorized as indecent, incompetent, perverse, irrelevant, backward, exotic or particular, or not perceived at all (Stalder 2016, p. 22 f.).

In the course of the (German) 1968 movement, a struggle for publicity has also been waged. Initially, leaflets were used as a form of articulation and demonstrated against the Springer press. At the same time, the 1968 movement also developed its impact through its media presence. The emergence of the student protests ran parallel to the development of the German television landscape: In 1963, ZDF[7] went on air, and by 1965 four more so-called third programs had followed. Entirely in keeping with the electronic age's implosion of space, the protest images disseminated by the mass media were close to the viewers via television. From this perspective, television seems to affect the media omnipresence of protesting Egyptians in the so-called Facebook Revolution of 2011.

Through television, every recipient can experience the protests through the immediacy of the images. But the suggestive effect of the images disseminated by the mass media remains trapped in the unidirectional structure that distinguishes television. The 'television makers' had the sovereignty of selection/interpretation over the images that were sent to recipients. This unidirectional structure also led to a media critique of the protesters of the 1968 movement:

> The protesters perceived television as an institution of domination and the reporting as directed against themselves. The student groups felt marginalized by the media and complained that their political content and demands were either ignored, defamed, or

---

[7]ZDF is short for Zweites Deutsches Fernsehen (Second German Television) is a German public-service television broadcaster based in Mainz, Rhineland-Palatinate. It is an independent nonprofit institution, which was founded by all federal states of Germany.

played down as riots by the students. […] As a consequence of their strict rejection of the public sphere existing in the Federal Republic, the student movement developed a counter-proposal. The concept of a counter-public sphere that it chose for this made the objective clear: the protesters had no intention of participating in or influencing the existing public sphere. They were more interested in establishing a new, independent public sphere that would create free space in the future (Vogel 2013, p. 68).

Stalder (2016) points out that the "freedom-oriented social movements of the 1960s began to perceive the mass media as part of the political system they were fighting against" (Stalder 2016, p. 70). This approach toward the counter-public sphere also found its echo in the media-theoretical considerations of its time. This aspect will be developed further in the following.

From a media theory perspective, Angermüller (2013) points out that the "mass media […] permeate everyday life in such a way" (Angermüller 2013, p. 249) "that they produce their objects of representation." The mass media evoke a protest culture, which remains in the ruling system's sphere solely because of its media structure. The counter-public sphere concept goes hand in hand with the claim to produce self-determined the 'images' of one's protest, or rather, not to have the voice of the protest lent by hegemonic mass media. With the formulation of the concept of the counter-public sphere, a critical attitude toward mass media was established in the electronic age. This approach influenced the subcultures of the digital age. In other words: a genealogy of media criticism can be reconstructed, ranging from the media criticism of the anti-authoritarian movements of the 1960s[8] to hacker groups such as Anonymous and platforms for a counter-public sphere such as Wikileaks.

With reference to the critical-doubting dimension of postmodern epistemology, the counter-public sphere model is used to create an approach for a space in which petits récits can set themselves apart from the hegemonic, metanarrative 'images'. The counter-public sphere functions as a free space that enables articulating postmodern protests. The establishment of free spaces through the unidirectional structure of the electronic age's mass media is reaching its limits. The critique of the media structure of the public sphere by actors in the protest movements of the 1960s and 1970s finds its media theoretical counterpart in the "Requiem for the Media" formulated by Baudrillard in 1972. Baudrillard's starting point is the thesis that how the media are handled is a metonymy for hierarchical social structures: "Not as vehicles of a content, but through form and operation, the media induce a

---

[8]This media criticism, in turn, has its precursors: in the German-speaking world, for example, media theoretical considerations of the Frankfurt School were taken up and specifically interpreted in the course of the 1968 movements.

social relationship" (Baudrillard 1978, p. 90). Baudrillard's Requiem for the Media is not unaffected by the political movements of the time:

> If one considers Baudrillard's historical and geographical position, it becomes clear that political events strongly influence France's theoretical orientation. Especially, the political and social events of 1968 in Nanterre dominate his academic life and influence his choice of topics (Yeh 2013, p. 307).

As in Western Germany and North America, the 1960s and 1970s represent "a time of upheaval" (Yeh 2013, p. 307). Yeh points out that the "mass demonstrations by American and West German student movements at the beginning of the year became the model for French student groups" (Yeh 2013, p. 307, footnote). Thus, "[i] n particular, the student protests of May 1968 in France [...] are a decisive historical and political turning point for Baudrillard" (Yeh 2013, p. 307 f.). These student protests culminated "in the occupation of administrative buildings, the closure of the university, violent conflicts with the police, and the arrest and imprisonment of students, which also provoked a general outrage among the population" (Yeh 2013, p. 308).

These social conflicts influenced Baudrillard's media theory which he developed in "Requiem for the Media." Baudrillard "deals with the mass protests and the role of the media in the context of the May riots in detail in 'Requiem for the Media'." (Yeh 2013, p. 308). Baudrillard states that "[m]an [...] could object" (Baudrillard 1978, p. 95) that "the mass media [...] in May '68 certainly played a role" (ibid.). That mass media like radio "spontaneously strengthened the revolutionary movement. At least for a moment, they (involuntarily) turned against power" (ibid.). From such a perspective, "[a]ll seems to speak in favor of believing in a subversive influence of the media during that time. Radios in French-speaking foreign countries and newspapers echoed the student action everywhere" (Baudrillard 1978, p. 95 f.). Baudrillard does not share such a perspective. Baudrillard does not understand the media attention of the student protests as the possibility of a (counter)public. He understands the media publicity that reports on the student protest as a media reproduction of the non-democratic power relations against which the students of May 1968 revolted in protest. From this perspective, Baudrillard's reflections on media theory, which he formulates in his "battle script" (Engell 2012, p. 72), have implications that are critical towards power: the unidirectional structure of television metonymically represents the authoritarian social relations of power. According to Baudrillard – "who can justifiably be regarded as a pioneer of the postmodernist discussion" (Angermüller 2013, p. 149) – the "mass media [...] are characterized by the fact that they are anti-mediatorial, intransitive,

in that they fabricate non-communication" (Baudrillard 1978, p. 91). In the course
of his reflections on media theory, Baudrillard critically dissociates himself from
Enzensberger's "Construction kit for a theory of media," which in turn was influenced by Bertolt Brecht's radio theory. In 1970, Enzensberger pleaded for the receiver to give himself a voice through media appropriation and thus raise himself
to a transmitter's status. In this way, a media practice is realized that undermines a
hegemonic-ideological effect of mass media. Instead of a transmitter reaching a
mass of individuals, a mass of individuals sends (revolutionary) content. According
to Baudrillard, such a construction, in which the receiver becomes the transmitter,
would merely mean a shift of emphasis within the unidirectional structure of
(mass) media: Another actor 'sends' content to other 'recipients.' A dialogical dynamic, is not realized. Therefore, a central thesis of Baudrillard is to understand
"communication as exchange" (Baudrillard 1978, p. 91). Communication is defined "as the reciproc[al] space of question and answer" (ibid.). Such a dialogical
structure cannot be realized through the unidirectional structure of mass media:
"[T]he media is that which forever prohibits the answer" (ibid.). If this thought is
continued, mass media of the electronic age, such as radio and television, do not
allow communication in the sense of dialogical exchange. Social structures of
power manifest themselves metonymically in the access to media: Thus, "power
[...] belongs to the one who can give and *to whom it cannot be returned*. To give,
and to give in such a way that one cannot be returned, that is, to break through the
exchange for one's advantage and establish a monopoly" (Baudrillard 1978, p. 91,
ed.). According to this – and this is a decisive difference to Enzensberger – for
Baudrillard, it is not sufficient that everyone owns media in order to initiate a democratization process:

> As if the possession of a television set or a camera opened up a new possibility of
> movement and exchange. But probably no more than the control of an icebox or a
> toaster. There can be no *answer* to a functional object: its function is to be an inte-
> grated speech that has already been answered, complied with and leaves no room for
> a game with a reciprocal *stake* (Baudrillard 1978, p. 93 f., e.i.o.).

Instead, communication from this perspective requires a "reciprocal space of
speech and *response*" (Baudrillard 1978, p. 91, ed.). However, "the entire contemporary architecture of the media [...] is based on this latter definition: *the media is
that which forever prohibits the response*, that which makes any process of
exchange impossible" (ibid., ed.). From this perspective, the electronic age's mass
media do not enable dialogical communication and, therefore, cannot structurally
establish the dialogical communication relationship that characterizes postmodern
communication. "It is, therefore, a strategic illusion to believe in the critical *use* of

the media. Today, such a speech is only conceivable through *destruction* as non-communication" (Baudrillard 1978, p. 101, ed.). As long as communication is understood as a transmitter-receiver relation, there can be no exchange since someone always sends actively and another receives passively consuming. It is not sufficient that revolutionary content is transmitted since it reproduces the established power structure on the media level. Through such media use, the "category of the 'transmitter' is preserved" (Baudrillard 1978, p. 109). Even if everyone had the possibility to broadcast, the absence of dialogical exchange "does not call the system of mass media into question" (ibid.).

*Postmodern communication is based on the epistemological perspective that in the dialogical relationship between petits récits, truth claims are examined epistemologically. This epistemological attitude consequently constitutes a subversive diversity that undermines the hegemonic claims of grand narratives. With reference to Baudrillard, it can be concluded that a specific medial dimension corresponds to the postmodern attitude of consciousness. This medial dimension must enable a dialogical relationship and overcome the mass media forms of communication of the electronic age.* This is an aspect that Yeh (2013) also points out: 'It is no longer the modern notion of reliable observability of social change, but the postmodern notion of unobservability or the fragmentation and multiplication of observer perspectives that becomes relevant through the dimension of media concepts' (Yeh 2013, p. 86). Following Yeh, the question can be formulated whether thinking about a postmodern communication constellation does not necessarily amount to thinking about media (see Yeh 2013, p. 86)? *Following on from this consideration and with reference to Baudrillard's media reflection, the discourses on Web 2.0 can also be read as the implicit discursive actualization of a dialogic-participatory communication in the sense of postmodern cognition: Through Web 2.0 and its participatory structure, a postmodern dialogical communication process is made possible – which has also produced specific protest practices organized via the Internet.*

When Baudrillard states that exchange or communication only becomes possible through the destruction of the medium (see Baudrillard 1978, p. 104), this consideration can be applied to Web 2.0: Web 2.0 media enable the destruction of the sender-receiver constellation of the electronic age. Through the dialogic, participatory orientation of Web 2.0 media, "the monopoly of speech is broken [...], not so that each individual can have the floor, but so that speech can be exchanged, given and returned" (Baudrillard 1978, p. 92). Against the background of these considerations, the position of or Downes (2005) can be understood, who sees the establishment of Web 2.0 as a 'social revolution': 'For all this technology, what is important to recognize is that the emergence of the Web 2.0 is not a technological revolution, it is a social revolution' (Downes 2005, para 2). The poly-directional and poly-phonic structure of Web 2.0 enables the democratization of communication on a medial level, which the mass media of the electronic age prohibit:

> Mass media such as radio establish a power relationship, their recipients are passively at the mercy of a *voice*. The communication here is one-sided. This asymmetrical communication is not communication in the true sense of the word. It resembles a proclamation. Therefore, such mass media have an affinity with power and domina-tion [...] Digital media, on the other hand, generate a genuinely communicative rela-tionship, that is, symmetrical communication. The receiver of the information is at the same time the sender. In this symmetrical communication space, it is difficult to in-stall power relations (Han 2013, p. 58, ed.).

In the following, we will work out how Web 2.0 media's participatory possibilities enable digitally based dialogical communication. This digitally-based dialogical communication requires new forms of postmodern protest *on* and *through the* Internet. Web 2.0 thus opens up the possibility of a postmodern protest culture of the digital.

## 2.2    Web 2.0 as a Mass Medium of the Digital Age

The term Web 2.0 has increasingly entered the discourse since 2003/2004. It is based on a changed perspective on the Internet, according to which users actively 'inscribe' themselves on the Internet: "Web 2.0 is the sum of efforts to make the Web more communicative and user-friendly" (Kantel 2009, p. 11). Powerfully, software developer Tim O'Reilly, who, among other things, helped develop the Perl scripting language and published the first book on the Internet in 1992, used the term in an article ("What is the Web 2.0?"). A year before the article was pub-lished, a so-called Web 2.0 conference was held for the first time in 2004, in which O'Reilly's publishing company O'Reilly Media was also involved. The term Web 2.0 began to establish itself and to have a discursive effect. Despite or perhaps be-cause of its discursive relevance, the term Web 2.0 is not without controversy. The term is sometimes accused of being conceptually vague and criticized as a buzz-word. This is also due to the proximity of the term to the technical language used in software development: In software development, numerical additions mark changes or further program developments. An increase before the decimal point signals a fundamental change compared to the previous version. Münker (2009) points out that there has never been such a suggested "version jump [...] on the Internet, however" (Münker 2009, p. 21). Instead, a "dynamic of the Internet[s] since its earliest days [lies] in the fact that these many programs and techniques are being researched and developed by just as many programmers and technicians around the world in processes that are independent of one another" (Münker 2009, p. 21). In the course of this decentralized and constant debate as well as the development of the technical potential of the Internet, it changes "continuously, not discretely" (Münker 2009, p. 21). These changes cannot be reduced to technological changes and affect the discursive interpretation of the Internet as a social phenomenon. Accordingly, Lehr (2012) elaborates that the "central characteristics" (Lehr 2012, p. 48) marked by the term Web 2.0 are "not of a purely technical nature" (ibid.). Instead, "the interaction of several social and technological currents and tools [...] brings about a change in the basic conception of the Internet" (Lehr 2012, p. 48). Web 2.0 describes not so much an entirely new technical structure as a specific form of interactive and collaborative use of the Internet (and here, primarily the World Wide Web). This changed form of Internet use can be worked out by contrasting a retrospectively constructed Web 1.0 with Web 2.0: The concept of Web 1.0 can be defined in the context of such a contrast by the fact that only a few people have access to the content and can edit the website. At the same time, many users can access this content. Consequently, Web 1.0 still exhibits the logic of a traditional understanding of mass media in the electronic age, where a medium addresses a mass of consumers: "As long as the Web could only be read, very many offerings were more related to the analog mass media of newspaper and television culture than one might initially suspect" (Münker 2009, p. 17). Web 2.0 gives the Internet a performative dimension of reality construction: "Most media environments in Web 2.0 only exist when we update them – and only because we use them" (Münker 2009, p. 10). Paradigmatically, this is demonstrated by so-called social networking sites (SNS) such as Facebook, Google+, or Snapchat. It should be noted that "[t]he Web 2.0 [...] does not replace the Web 1.0" (Münker 2009, p. 80) – instead, it "steps alongside it" (ibid.).

With Web 2.0, a redefinition of the concept of mass media becomes virulent. While television as a mass medium of the electronic age was unidirectional through a one-to-many structure, a dialogic mass medium is established at the latest with the Internet's many-to-many structure. With the establishment of the concept of Web 2.0, the Internet becomes the mass medium of the digital age: Potentially, masses of senders can reach masses of receivers. Simultaneously, these senders can react dialogically to the content – for example, via a comment function of a weblog, in the joint discussion of wiki contributions, etc.

Consequently, Web 2.0 does not mean an update or a new version of the Internet. Rather, an understanding of the Internet has been discursively established that offers the media basis for a collaborative generation of knowledge in the poly-directional and polyphonic functions of web tools such as wikis, blogs, and social bookmarking applications. In the course of this, the perspective on the Internet is changed. Discursively, "[t]he net [...] is reinterpreted from a supply surface to an application environment" (Münker 2009, p. 21). Crucial to this is the changed role of users, "who are no longer merely demanders of a service but actively contribute to its creation" (Lehr 2012, p. 48). The participatory dimension that defines the cultures of the digital finds its mass media infrastructure in Web 2.0. The changes marked by the term Web 2.0 restructure social reality, redefine social figurations, and consequently lead to new forms of self/world relations.

Epistemological analyses represent a heuristic starting point for elaborating these effects of media shifts. Through these analyses, the relationship between the individual and the (surrounding) world can be reconstructed in the medial change context leading to a digital age. Within these analyses' framework, new forms of digitally-based knowledge generation and digitally-based protest can be taken into an analytical view on an epistemological basis.

## 2.2.1  Epistemological Perspective on the Subject in Medial Change

As a field of research in philosophy, epistemology asks about the possibilities and limits of cognition. A central focus of epistemological analyses lies in the relationship between the cognizing subject and the world to be cognized. This relationship can also be understood as a basic epistemological constellation. In the course of the fundamental epistemological constellation analysis, it is also necessary to work out who or what is the cognitive process agent.

In modern epistemology, Descartes, "whose name has come to symbolize the modern rational worldview" (Hartmann 2000, p. 36), identifies the critical cogito as the agent of cognition. Cognition is the result of a strategic doubt: the subject cognizes itself by questioning all knowledge. The ego is installed as the legitimate cognizing agency. The Cartesian subject is autoreferential and thus corresponds to the non-dialogical, unidirectional structure of the book. Hartmann (2000) notes that in the course of Descartes' epistemological research, the 'keywords' "codification, universalization, globalization [...] can be used to circumscribe the presupposition of that philosophical thinking" (Hartmann 2000, p. 43) "which accepts as truth only the results of abstract thinking based on what can be 'clearly and distinctly known. But these are also presuppositions that were already based on a functioning foundation for intellectual activity: *book culture*" (Hartmann 2000,

p. 43, ed.). It is in the inward, reflexive self-reference that cognition manifests itself. With autoreferentiality as a form of valid cognition, the rationally thinking and acting citizen is epistemologically established as the central actor of bourgeois society. Adorno and Horkheimer also address this connection between social self-understanding and epistemological analyses when they state that "[t]he individuals who have to take care of themselves [...] develop the ego as the instance of reflective foresight and overview" (Adorno and Horkheimer 1997, p. 106).

In reading and writing, cultural techniques have constituted that base on the media structure of the Gutenberg Galaxy and medially prefigure forms of the active bourgeois individual: "On the one hand, reading and writing practices form [...] an autonomous field of education and introspection, but at the same time they are closely linked to the activities of bourgeois work and also bourgeois personal relationships" (Reckwitz 2006, p. 98). Epistemological analyses, from this perspective, are also effects of the social contexts from which they originate. As the understanding of culture changes with social change, so does the conceptualization of the fundamental epistemological constellation of cognizing subject/world to be cognized. The analytical processing of the fundamental epistemological constellation is dependent on the medial structure of the world that is to be recognized. This aspect will be discussed in the following, based on the concept of 'the author.'

## 2.2.2   The Author as an Agent of Knowledge

The author's function can be used as a paradigm for tracing the shift in the generation of knowledge through medial change. The starting point is the thesis that the author represents a central feature of the book culture that emerged in the Gutenberg Galaxy course. With the book as the leading medium of the Gutenberg Galaxy, the author as the creator of the book also becomes relevant: "If the thinker were not at the same time and literally in the act of thinking an author, his project would forfeit the very contingency he sought to overcome"'(Hartmann 2000, p. 44). Through the book, the written fixation of language was 'inscribed' in the cultural landscape of modern Europe, or rather, the media foundation was laid for the dawning bourgeois age: "The printing press was [...] the *new* reality" (McLuhan 1968, p. 290, ed.). In this context, the printing press's establishment conditioned a "shift from an oral to a visual culture" (McLuhan 1968, p. 314). Bourgeois society represented itself through books and fought out its discourses of self-understanding (see Kergel 2011a, b). This also implies the existence of an academic, public sphere that is constituted, among other things, through book culture:

In the end, the experience of the world only serves the experience of the self, but under conditions of a scientific publication system that is already functioning in rudimentary form, without which private thoughts would be nothing more than that, and could quite impossibly serve the re-foundation of a scientific method (Hartmann 2000, p. 44).

The historical significance of Gutenberg's process can be seen in the fact that letterpress printing with movable metal type and the printing press initiated a profound medial change towards book culture. For example, Stalder points out that "in the first two generations after Gutenberg invented modern letterpress printing, i.e., between 1450 and 1500 [...] more books [were] produced than in the 1000 years before" (Stalder 2016, p. 102). The Gutenberg process also made it possible to produce larger written works inexpensively in higher print runs. This set the stage for a broad book market. With the spread of Gutenberg technology, the book increasingly became the leading medium and caused profound cultural changes, which McLuhan theorized with the Gutenberg Galaxy concept.

According to McLuhan's analyses, book culture represents the medial basis and contributes decisively to the supra-regional and broad establishment of bourgeois discourses of self-understanding and cultural identity construction:

> If between the sixteenth and eighteenth centuries a modern subject emerges as a bourgeois subject within the framework of a specifically bourgeois practice, then the media practices in dealing with the written word in the technological form of book printing, which first made possible the supra-regional dissemination of written material, i.e., a specifically bourgeois-modern form of reading, and besides also of writing, present themselves as the practice goal of a specifically bourgeois habitus (Reckwitz 2006, p. 98).

The individualism that characterized bourgeois society (see Kergel 2013) and established itself based on a bourgeois reading culture can be seen metonymically in the author's role: book printing opened up a media forum for the bourgeois individual to articulate. With the book, the author became relevant as the instance of knowledge and at the same time as the author of the book.

In his text "The Death of the Author" (1968), published 6 years after McLuhan's "Gutenberg Galaxy," Barthes elaborates on how the figure of the author shapes 'our culture': "Our contemporary culture tyrannically confines literature to the author, to his person, his story his passions" (Barthes 2016, p. 186). In writing, the author represents the figure of the productive/creative citizen – he is a Man of Letters.

In his essay "What is an Author," which appeared a year after the publication of Barthes' text, Foucault states that the individual is discursively grasped as a producing "being of reason" (Foucault 2016, p. 214). Cognition, experience, creative power is displaced to the interior according to Foucault's analyses. Thus, in the "individual [...] there is supposed to be a 'deep' urge, creative power, a 'draft' and this is supposed to be the place of origin of writing" (Foucault 2016, p. 214). Foucault locates the figure of the author as an effect of discursive self-identification of bourgeois culture. Accordingly, the author is part of "industrial and bourgeois society, individualism and private property" (Foucault 2016, p. 228). The relevance to the cultural discourses of self-understanding of bourgeois society effects a cross-field impact of the figure of the author. Thus, the meaning of the author becomes virulent in the legal field, among others. The introduction of copyright is an effect of the author and, therefore, one of the Gutenberg Galaxy features. The creative individual produces ownership with his text:

> By the middle of the 19th century, the national regulation of the author's right had already been completed (in the USA in 1790, in France in 1793, in Prussia in 1837, and then in the German Empire in 1870). These laws were the result of an interplay of aesthetic and legal discourses in the 18th century, which placed the "author" at the center of new legislation and granted him economic and, in part, moral rights to his work for a limited period (Dommann 2008, p. 44).

In addition to the legal dimension, the figure of the author is also actualized in epistemological discourses. As an epistemic agent, the author represents the autoreferential knowledge that Descartes founded epistemologically with strategic doubt. The author is primarily *not a* collective of authors but *an* individual. From the first person plural, (world) knowledge is reported in written language. Written language becomes the medium of cognition, the cognizing subject of the author. Descartes' inauguration of the cogito epistemologically anticipates the thesis of the linguistic turn that world experience and world knowledge unfold through and in language. Davenport (2006) accordingly establishes a connection between Austin's speech act theory and Descartes' linguistically manifested positing of the self (see Davenport 2006):

> Descartes' argument depends on the assumption that human speech acts imply agent causation. The self-referential speech act 'I exist' strikes us irrefutable only because the first-person represented in the speech-act is assumed to be the causal agent without whom the speech-act would not be performed. In short, we assume that it is authentically self-referential (Davenport 2006, p. 132 f.).

The subject, which grasps itself self-reflexively in the forum of language, can verbalize and analyze impressions/experiences through self-reflection. Through this, the subject can relate to these impressions/experiences in an epistemic-critical way. The epistemic, autoreferential dimension of cognition is thereby medially fed back. Through writing, the philosopher as author objectifies his insights based on doubt in the book. Through this writing objectification, the inwardness of the philosopher as an auto-referential author is established. Cognition is generated in terms of written language: "The early, bourgeois form of modernity with its specifically bourgeois form of the subject, the inward-oriented cognitive-moral-emotional orientation of subjectivity, is based on the written culture in the form of book printing" (Reckwitz 2006, p. 92). (Written) language as a metonymic representation of objectified interiority becomes the central medium of cognition. "If, for example, one states an autobiographical consciousness of the subject in bourgeois culture, then the praxeological perspective asks about the practices of diary writing as an activity that promotes such autobiographical interiority" (Reckwitz 2006, p. 95). Here, the book is the medium in which language materializes in terms of written language: "the bourgeois subject of writing appropriates dispositions that are constitutive of all bourgeois practice: a cognitive concentration of attention, a psychologization, and affective sensitization, finally a moral sense" (Reckwitz 2006, p. 98). Through the engagement with written language, "the bourgeois subject [...] emerges as an inwardly oriented one, as a subject with a complex inner world" (ibid.). This 'complex inner world,' which finds an epistemological-aesthetic program in, among other things, Romantic, expressive aesthetics, is "populated by cognitive as well as moral and emotional elements" (ibid.). The bourgeois subject reveals itself through writing and manifests itself in the author's discursive figure as a bourgeois instance of knowledge. With the role of the author, with the book as the leading medium of the Gutenberg Galaxy, and the concept of the author, a structure of domination is constituted medially between author and reader or producer and recipient. The creative author creates the text, his 'Eigentum.' This 'Eigen-Tum' is made accessible to the reader via the infrastructure of book culture: "The book as such can only be printed [...] thus become part of communication because there are publishers who print and circulate books, as well as bookshops and libraries through which the books are distributed. This is the technical pole of the medium" (Schwalbe 2011, p. 130). The unidirectional orientation of the book separates the author and the reader in a certain way. "The author strives to break out of the self-sufficiency of subjective reason; in other words, he wants to be read" (Hartmann 2000, p. 45). The separation of roles between author and reader also manifests itself in the scientific field. The scientific author communicates knowledge findings to the recipient through publications, which the recipient absorbs by reading. The image of the student sitting in front of the book has acquired an iconic significance. The book becomes a central medium of knowledge; the world is appropriated by reading. Hartmann (2000) points out that with McLuhan's media theoretical reflection on the electronic age, the Cartesian subject is also challenged. In the course of the medial change, the medial conditions of cognition also change and thus also the constitution and self-assurance processes of the subject: the Cartesian epistemological constellation is subject to a transformation process with the medial change: With the poly-directional and polyphonic possibilities of Web 2.0, the separation between author and recipient is eroding. Web 2.0 media such as collaborative writing programs or wikis enable dialogic-collaborative forms of cognition and knowledge generation. Through the dialogical potentials of Web 2.0, the individual author's autoreferential concept as an agent of knowledge is replaced by collective authorship.

### 2.2.3   Rhizomorphic Cognition – Root Network Instead of the Tree Structure

Cognitive processes *with* which and *through* which the subject emerges as a cognitive instance change in the course of medial transformations. This also changes subject configurations or how the subject generates self/world relations. In the digital age, an implosion of time and space is medially produced through digital media, which leads to a constitution of 'real-time' or evokes presence:

> Under the historical auspices of high-performance computers [...] and data links [...] that ideally collect, store, calculate and transmit information at that speed of light [...] that appears to us as 'real-time,' it has become available everywhere simultaneously. With this implosion of space and time of information in computers, our concepts of space and time have simultaneously exploded [...] Where every point in cyberspace can be reached simultaneously from every other point, the simultaneity of the non-simultaneous and the omnipresence of the dispersed shape our perception (Stingelin 2000, p. 17).

The implosion of time and space, which has already been prepared or accomplished by the image world of the electronic age, seems to be radicalized by the Internet and expanded by the participatory dimension of the digital's cultures in action in a production-oriented way. Suppose the Internet medially expands forms of self-perception/world-perception. In that case, it must be asked from an epistemological perspective how the basic epistemological constellation of subject/world to be known can be appropriately conceptualized against the background of the medial transformation of social reality?

In the 1960s and 1970s, i.e., at a time when the foundations of the Internet were being laid, and protest movements were forming, a power-analytical understanding of the cognizing subject was developed in the context of poststructuralist epistemology, among other things: The Cartesian figure of the self-referential cognizing subject comes into crisis at the latest with poststructuralist concepts of the subject. For example, Foucault deciphers the subject as an epistemological construction of the bourgeois age and as an effect of power constellations or processes of subjectivation. Another starting point for shifting the perspective on the basic epistemological constellation is the rhizome model developed by Deleuze and Guattari in the context of the protest movements of the 1960s and 1970s. As an epistemological approach, the rhizome model substitutes the autoreferential cogito in favor of a collaborative form of cognition. Rather than through an epistemic, doubting view of the world, the world is known or constructed through social interaction. The cognizing ego dissolves into networks of knowledge characterized by multiple perspectives. Cognition becomes a collaborative process in the course of which individuals align their experiences of self/world. In this process of alignment, the distinct boundaries between cognizing subjects disappear: The individual cogito merges into a multiplicity of cognizers. In this way, Deleuze and Guattari formulate a 'net-like' strategy of cognition organized in a decentralized manner analogous to the hypertext structure. To capture this structure of cognition analytically, Deleuze and Guattari resort to the rhizome as a metaphor for a form of decentralized knowledge organization. The authors thus formulated an approach that "[i]n the French as well as in the German subculture triggered a veritable *rhizomania*" (Hartmann 2000, p. 299, ed.) and "has been formative for what can be called the net discourse to this day" (ibid.). Ott (2005) points out that the epistemological reorientation carried out by Deleuze, together with Guattari, is also shaped by the "events of May 1968". "The political events" entailed a "stronge[r] involvement of the political field" (Ott 2005, p. 97).

Consequently, the critique of taxonomic orders and the counter-model of the rhizome also have inherent power-analytical and domination-critical implications. When Baudrillard notes that social power structures are also metonymically manifested in the organization of media or in the way media are used, an analogous consideration of Deleuze and Guattari is actualized in the field of epistemological reflection: The rhizome represents a form of organization that, read from a sociocritical perspective, opposes centralizing forms of social organization with the model of a decentralized, anarchic-associative form of organization. This critique of centralizing forms of organization can also be seen in choosing the metaphor of the rhizome. Deleuze and Guattari borrow the word rhizome from biology, where it denotes a root network. Here, "root and shoot are indistinguishable" (Hartmann 2000, p. 301). Hierarchies and taxonomic orders do not determine the root network, but by structural flexibility: "It is perhaps one of the most important properties of the rhizome to always offer multiple means of access" (Deleuze and Guattari

1992, p. 24). These considerations on a rhizomatic organizational structure can also be applied to the process of knowledge organization: rhizomatically structured knowledge is constructed by relating information to each other. The process of relating does *not* generate a final hierarchical or taxonomic knowledge order. Instead, the assignments are part of a flexible process. Information can always be placed in new contexts, generating new knowledge: "A rhizome can be interrupted or torn at any point. It continues along its own or other lines. One cannot cope with ants because they form an animal rhizome that continues to form even when its largest part is destroyed" (Deleuze and Guattari 1992, p. 19).

Rhizomatic knowledge is ephemeral knowledge in constant becoming: "Every rhizome contains segmentation lines that stratify, territorialize, organize, designate, assign, etc.; but also deterritorialization lines that allow escape at any time" (Deleuze and Guattari 1992, p. 19). To elaborate analytically on this form of rhizomatic knowledge generation or knowledge organization, Deleuze and Guattari distinguish the rhizome model from the knowledge tree model. As a metaphor, the knowledge tree refers to taxonomic forms of knowledge organization that can be found, for example, in family trees that depict relationships of kinship (Figs. 2.3 and 2.4).

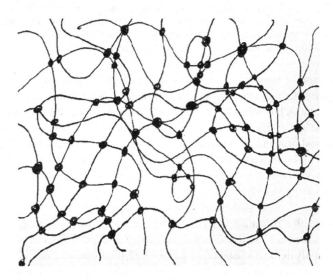

**Fig. 2.3** Visualization of a rhizome structure. Stable taxonomic orders are made impossible by cross-links that are in motion (own drawing)

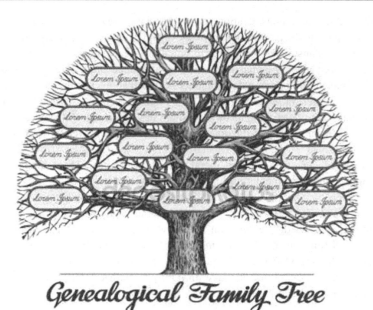

**Fig. 2.4** The branches symbolize the categories. The branching of a tree represents in this metaphor the categories and subcategories of a taxonomic order. (Source: https://st2.depositphotos.com/1496387/11229/v/950/depositphotos_112290586-stock-illustration-vintage-genealogical-family-tree-leafless.jpg, last accessed October 14, 2017)

Significant for knowledge trees are classifications characterized by categories and subcategories and thus constitute a knowledge order. Classification is carried out based on a dichotomous distinction. Each element is assigned to an order level or a position in the knowledge tree. An element cannot have several positions in the organizational structure of a knowledge tree. Suppose it is assumed that knowledge organization processes also performatively construct or reproduce an understanding of society. In that case, the knowledge tree can be read as a metaphor for a hierarchically defined understanding of the world/society. Rhizomatic perspectives subversively undermine such an understanding. Thus, a rhizomatic form of knowledge organization is characterized by the fact that manifold positionings or cross-connections are possible. Such "cross-connections between differentiated lineages upset the family trees" (Deleuze and Guattari 1992, p. 21), orders erode. Concerning epistemologically oriented considerations of postmodern disorder, rhizomatic cognitive processes can be conceived as postmodern cognitive processes of disorder.

The rhizome deconstructs hierarchical orders, which can also be read as ordering models of metanarratives. From this perspective, metanarratives are coherent entities, which require a totalizing taxonomically organized order of things. If this thought is taken further, the consequence is that every element has its place in the overall structure of the metanarrative. Metanarratives do not admit anything foreign but make the unfamiliar its own and assign it a place in its structure of meaning (this subsumption logic is also a feature of integration discourses).

Against the background of postmodern epistemology, the rhizome model can be understood as a counter-design to the unifying totalizing tendencies of metanarratives: Just as the plurality of petits récits, gives plural access to experiences of reality or allows for different forms of reality construction, the rhizome describes a model of knowledge organization that is also characterized via a plurality of knowledge accesses. Accordingly, "perhaps one of the most important properties of the rhizome is always to offer multiple access points" (Deleuze and Guattari 1992, p. 24). These considerations of multiple means of access are also followed by the organization of the book "A Thousand Plateaus," within which Guattari and Deleuze unfold their rhizomatic model of knowledge in 1980. In this way, Deleuze and Guattari abandon the concept of the singular author. In this context, the rhizome stands as a metaphor for a collaborative, non-hierarchical, and non-taxonomic form of cognition: if the tree model represents the static of the hierarchical form of knowledge organization, the inconclusiveness as well as structural openness of the rhizome map an infinite dynamic of collaborative knowledge generation processes: "In rhizomes there are tree and root structures, but conversely, the branch of a tree or the part of a root can also begin to sprout rhizome-like buds" (Deleuze and Guattari 1992, p. 27). Deleuze and Guattari extend this rhizomatic form of knowledge generation to an attitude of knowledge, which they capture with the adjective 'rhizomorphic':

> To be rhizomorphic is to produce strands and fibers that look like roots, or instead join with them by penetrating the trunk, even at the risk of some new and unusual use being made of them. We have grown weary of the tree. We must no longer believe in trees, in roots tremendous and small; we have suffered too much from them. The whole arboreal culture is based on them, from biology to linguistics. Only underground strands and aerial roots, the wild growth and the rhizome, are beautiful, political, and loving (Deleuze and Guattari 1992, p. 27).

According to Deleuze and Guattari, the rhizomorphic attitude to knowledge deconstructs the linear structures of the book. Thus, Deleuze and Guattari also understand their book "A Thousand Plateaus" not as homogeneous but as rhizomatically structured. This structure is also reflected in the writing process, which is not written by an author, but by a 'we': "We write this book like a rhizome" (Deleuze and Guattari 1992, p. 37). Instead of a through-composed structure – "A book [...] made up of chapters has its climaxes and conclusions" (Deleuze and Guattari 1992, p. 37) – Deleuze and Guattari write a book made up of 'plateaus.' Here Deleuze and Guattari refer to "any manifold as a 'plateau' that can be connected to other manifolds by extremely fine subterranean strands so that a rhizome can emerge and spread" (Deleuze and Guattari 1992, p. 37). In terms of a content-form linkage, the book "A Thousand Plateaus" is consequently conceived so that it can be received non-linearly. Simultaneously, the book is open to cognition; the construction of meaning is left to the reader. "A book has neither an object nor a subject; it is made up of various-shaped matters, of the most diverse data and speeds" (Deleuze and Guattari 1992, p. 13). This redefines the author/reader power relation. The author does not provide a linear, completed chain of argument. Instead, it is up to the reader to give the book a structure of meaning. This decentralized structure and decentering of the author corresponds to a form of non-linear or rhizomorphic re-ception: "Each plateau can be read from any point and related to any other" (Deleuze and Guattari 1992, p. 37). This form of the book, where the reader can be 'immersed in the book at any point, breaks with a "discursive linearity" (Hartmann 2000, p. 299) that characterizes the reading structure of the Guttenberg Galaxy. In such a non-linear understanding of the book's structure, the reader represents the stepping point that first constitutes the book in the course of its construction through an interrelation of plateaus. Following this enthronement of the reader, the rhizomorphic book is conceived as open to interpretation: "As with all other things, in a book, there are also dividing or segmenting lines, layers or territories; but also lines of flight, movements that dissolve territorialization and stratification" (Deleuze and Guattari 1992, p. 13).

Deleuze and Guattari write a rhizomatic book that can be read as a consequence of Barthes' reflections on the 'death of the author': Barthes deconstructs the au-thor/reader relation by defining the reader as the very constituent of the text, thus subversively undermining the 'domination of the *author*' (Barthes 2016, p. 191, ed.):

> The reader is the space in which all the quotations that make up a writing are inscribed without a single one being lost. The unity of a text lies not in its origin. Still, in its destination – whereby this destination can no longer be understood as a person [...] He is only the *someone* who unites in a single field all the traces that makeup what is written (Barthes 2016, p. 192, ed.).

The text is first constituted as such by the recipient, who as a reader, charges it with meaning. In the process of this generation of meaning, other texts flow into it, forming the horizon of understanding, as it were, against which a new text is received and thus interpreted. The reader functions as a node of different texts. If the reader, in turn, writes a text himself, the background of understanding created by the received texts also flows into it. A text thus contains intertextual references,

- which are placed in the developing text by the author in the process of writing, and
- which are placed in the text by the reader in the process of reception.

One consequence of this double intertextuality is that the text is not a closed unit of meaning but can be reconstructed as a rhizome structure. The intersections or the nodes are the readers who constitute a text in the act of reception. The book as an object dissolves into rhizomatic structures: "That is why a book has no object. As a structure, it itself exists only in connections with other structures" (Deleuze and Guattari 1992, p. 13). The book's infinity, which Deleuze and Guattari develop epistemologically and Barthes elaborates on in reception theory, finds its medial equivalent in the so-called 'remix and mash-up culture.' This 'remix and mash-up culture' is establishing in the course of the communicative possibilities of Web 2.0 and finds a suitable licensing model in the Creative Common approach.

The reader becomes the author, continues to write the book, and thus rewrites it. The book as a self-contained knowledge entity erodes in the ephemeral dialogical structure of Web 2.0. The dissolution of the Cartesian cogito accompanies the dissolution of the book's linear structure in favor of collaborative cognition: social practice detaches itself from a fixation of the individual. This also erodes the author and reader's ontological difference: "The tree is filiation, but the rhizome is alliance" (Deleuze and Guattari 1992, p. 41). The Cartesian auto-reflexive cogito merges into a collaborative attitude to knowledge. Raulet conceptualized this recasting of the epistemic subject as early as 1988 regarding the decentralized structure of 'new technologies': "The subject has long ceased to be the constitutive center of the space of observation; synthetic image production and, more fundamentally, communication networks have undermined its privileged position" (Raulet 1988, p. 180). Here, Raulet refers to the deconstruction of the subject in the context of poststructuralist epistemology. Regarding the questioning of the subject as an agent of cognition, he notes that "[v]irtually twenty years ago [...] one would have concluded that the *structure* is the subject. The conclusion still holds, but it must be sharpened to the foreseeable effects of the new information or communication technologies" (Raulet 1988, p. 180, ed.). Raulet formulates a cognitive dynamic that corresponds on an epistemological level to the instantaneity of the hyper-culture formulated by Han: "The interconnectedness and the general interchangeability that dissolve local ties allow them only in the form of momentary, floating identities, isolated and multiplied language games, unpredictable encounters" (Raulet 1988, p. 180). What Raulet describes here can also be read as the ephemeral structure of a decentered, rhizomorphic attitude to knowledge. In this reading, encounters represent nodes that facilitate a dialogic-collaborative form of knowledge production. Accordingly, Deleuze and Guattari speak of "collective[s] fabric[s] of enunciation" (Deleuze and Guattari 1992, p. 38). This collective structure of enunciation finds its infrastructural counterpart in the model of the hypertext.

## 2.2.4 Hypertext and Memex – Structures of Non-linear Thinking

The poststructuralist upgrading of the reader from passive knowledge consumer to active knowledge producer finds a media infrastructure through Web 2.0. The potentials of Web 2.0 make it possible to transcend the unidirectional relationship of domination of book culture towards a participatory dialogue. This process of change is also characterized by a loss of the linearity established by book culture. The syntagmatic structure of written language constitutes a linear path of cognition of logical argumentation and materializes in the book, which is received linearly from left to right. The rhizomatic structure of text reception and knowledge production, on the other hand, finds its medial equivalent in the decentralized structure of the Internet. Stingelin (2000) pointed out the fit between Deleuze's and Guattari's epistemology and the decentralized structure of the Internet even before the concept of Web 2.0 was established:

> The rhizome poetics read retrospectively not only like an anticipation of the freedom of movement in space and time that is technically implemented in cyberspace [...] From this perspective, the Internet appears to be an epiphany of the philosophy of Deleuze and Guattari. In their self-understanding, the most philosophical of the cybernauts navigate the multifariously intertwined space/time continuum of cyberspace by orienting themselves with the help of concepts from Mille plateaux. Here, as there, 'the nearest and the furthest' combine ubiquity and simultaneity, ubiquity and simultaneity. Like the Internet, which is a network of networks, Mille plateaux exists as a structure (agencement) as a linkage, itself only in connection with other structures (Stingelin 2000, p. 19 f.).

The structural analogy between the rhizome and the many-to-many communication structure of the Internet can be paradigmatically established in hypertext.

Hypertext can be understood as a net-like structure that links information/objects with other (hypertext nodes) by through cross-references (hyperlinks): "'Hypertext' simply means that within a text document, a link is created with another document, which can be accessed via 'links' at a different storage location (similar to the classic footnote, which refers to content outside the present text)" (Hartmann 2006, p. 178). Hypertext allows for non-linear "jumps between references" (Kirpal and Vogel 2006, p. 143) that can be executed quickly via a computer. The content representation here is characterized by a multimodal interweaving of writing, images, video, and sound, typical of hypertext. "Computerized writing moves away from linearity in hyperlinks and search [...]" (Palm 2006, p. 126). There is also a semiotic dimension to the non-linear and multimodal structure of hypertext. Accordingly, hypertext becomes a metonymy of non-linear, associative thinking subtextually demarcated from the syntagmatic structure of book culture: 'Antilinearity is a subversive element of tension in the linear structure of writing' (Palm 2006, p. 126). Hypertext enables a rhizomatic-disruptive form of movement:

> We jump into the interstices of texts, rearrange temporal orientations, and convert the course of events in memory into structures that follow their assessed meaning rather than chronology. The hypertext, a text that advances to a complex network of references from others and itself, is radicalized in the digital medium (Palm 2006, p. 126).

The hypertext structure is given a semiotic dimension – analogous to Deleuze's and Guattari's rhizome model – that subversively undermines hierarchical structures of domination through the processual constitution of associations. The hypertext is not merely a structure but advances to become a symbol of non-taxonomic, non-hierarchical thinking. Thus, Nelson (1965) refers to the associative textual structure of hypertext:

> Let me introduce the word "hypertext" to mean a body of written or pictorial material interconnected in such a complex way that it could not conveniently be presented or represented on paper. It may contain summaries or maps of its contents and their interrelations; it may contain annotations, additions, and footnotes from scholars who have examined it (Nelson 1965, p. 96).

In 1960, Nelson founded the Xanadu project (named after the place where Kubla Khan had a pleasure palace built). Nelson's project envisaged a decentralized storage project in which documents were to be linked associatively in the sense of a universal library. This project failed due to its complexity. However, Nelson's attempt to adopt a technical infrastructure to the decentralized, non-linear structure of knowledge organization can be interpreted as an important conceptual step on the path leading from the electronic age to the digital age. Others have already prepared this path, Nelson's approach has a tradition: for example, Nelson's concept of hypertext has its roots in the 'Memex' (Memory Extended) model. This model was developed in 1945 in the article "As we may think" in the magazine "The Atlantic Monthly" by Vannevar Bush.

Memex represents a machine conceived by Bush with which knowledge can be collected and organized: "A memex is a device in which an individual stores all his books, records, and communications, and which is mechanized so that it may be consulted with exceeding speed and flexibility" (Bush 1945, p. 121). The Memex enables an associative form of knowledge organization: "It affords an immediate step, however, to associative indexing, the basic idea of which is a provision whereby any item may be caused at will to select immediately and automatically another" (Bush 1945, p. 123). According to Bush, this form of associative indexing corresponds to the human thought structure. Epistemologically, Bush, like Nelson and Deleuze and Guattari, adopts linear thinking, which can be understood with reference to McLuhan as a central feature of the book culture of the Gutenberg Galaxy:

> The real heart of the matter of selection, however, goes deeper than a lag in the adoption of mechanisms by libraries, or a lack of development of devices for their use. Our ineptitude in getting at the record is largely caused by the artificiality of systems of indexing. When data of any sort are placed in storage, they are filed alphabetically or numerically and information is found (when it is) by tracing it down from subclass to subclass. It can be in only one place, unless duplicates are used; one has to have rules as to which path will locate it, and the rules are cumbersome. Having found one item, moreover, one has to emerge from the system and re-enter on a new path. The human mind does not work that way. It operates by association. With one item in its grasp, it snaps instantly to the next that is suggested by the association of thoughts, in accordance with some intricate web of trails carried by the cells of the brain. It has other characteristics, of course; trails that are not frequently followed are prone to fade, items are not fully permanent, memory is transitory. Yet the speed of action, the intricacy of trails, and the details of mental pictures are awe-inspiring beyond all else in nature (Bush 1945, p. 121).

Regarding the state of research at the time, Bush designs the model of a Memex machine, which, as a form of analog computer, anticipates hypertext structures as thematized by Ted Nelson in the 1960s. In the process, a specific attitude of cognition is conceptualized that can be described, with recourse to Reckwitz, as a computer subject: "The computer subject now trains itself in the attitude of a 'user' who engages in a constant 'exploring': an unfinished, tentative search for aesthetic stimuli that is guided by the possibilities of association and combination" (Reckwitz 2006, p. 102). This computer subject is thereby interwoven with the ephemeral structure that characterizes the culture of the digital: "In the medium of the computer, the subject learns to produce simulations constantly and to be confronted with simulations that possess a provisionality and changeability and make different versions playable" (Reckwitz 2006, p. 103).

The decentralized hypertext structure of the Internet or the WWW resolves poststructuralist considerations on the non-linear cognitive process at the media level. Web 2.0, on the other hand, links this decentralized organizational structure with a participatory involvement of the user through the collaborative potential of participatory media such as wikis and blogs. Through Web 2.0 media, the user can create low-threshold web content and goes from being a mere consumer of web content to a producer or 'prosumer' (see Toffler 1980). This opening up of Internet communication opens up a dialogic space that overcomes the unidirectional orientation of media criticized by Baudrillard. The dialogical 'activation of the Internet user' finds its epistemological counterparts in the enthronement of the reader performed by Barthes as well as by Deleuze and Guattari's. The establishment of an associative and non-hierarchical understanding of cognition, which shifts the emphasis away from knowledge reception towards knowledge construction, necessitates a decentralized understanding of communication processes. Discursively, these considerations have been formulated in terms of media theory and epistemology, partly in the context of the anti-authoritarian movements of the 1960s and 1970s. In this context, postmodern figures of argumentation are updated on the level of media theory and epistemology: The subject gets into motion; cognition is rhizomorphically generated dialogically in the social context. In the course of this process of cognition, the hegemonic claims of metanarratives are subversively undermined. Such a conception of cognition erodes a traditional epistemological perspective. The basic epistemological constellation subject/world to be cognized is modeled in terms of a dichotomous subject/object relation. Instead, knowledge about the world is constructed in intersubjective, rhizomatic communication contexts. Forms of Web 2.0 supported knowledge construction in the digital age appear to enable such a form of knowledge. Thus, connectivism presents a model that reworks the decentralized, ephemeral knowledge construction processes in the digital age in learning theory.

## 2.2.5   Learning Culture in the Digital Age – Connectivism and e-Learning 2.0

The possibilities of digital-based interaction are also being discussed in the pedagogical field. Thus, it is necessary to work through the considerations on a form of

Web 2.0-based collaborative cognition in the digital age in terms of learning theory. This is paradigmatically done by Siemens (2004) with the approach of connectivism, which, according to Siemens is supposed to be a learning theory for the digital age:

> Behaviorism, cognitivism, and constructivism are the three broad learning theories most often utilized in the creation of instructional environments. These theories, however, were developed in a time when learning was not impacted through technology. Over the last twenty years, technology has reorganized how we live, how we communicate, and how we learn. Learning needs and theories that describe learning principles and processes, should be reflective of underlying social environments (Siemens 2004, para. 1).

From a learning theory perspective, connectivism makes it possible to describe interaction processes made possible by Web 2.0 media. Accordingly, connectivism focuses on the decentralized possibilities that Web 2.0 media applications open up for learning contexts. Grünewald et al. (2013) point to the connection of connectivism with Web 2.0 when they state that "[t]he philosophy of Web 2.0, which has given rise to the collaborative creation of, for example, environments such as wikis and forums, [...] supports diverse facets of connectivism" (Grünewald et al. 2013, p. 145).

According to the approach of connectivism, learning processes in the digital age are characterized by the fact that the learner has the opportunity to participate in different learning communities (e.g., in wikis, MOOCs, chat rooms, etc.) or to 'connect' to these learning communities: "In connectivism, learning occurs when a learner connects to a learning community and feeds information into it" (Şahin 2012, p. 442). Learning in the context of connectivism is self-directed learning. The "freedom of learners" allows them to "access the content through their own learning paths" (Grünewald et al. 2013, p. 144). The decentralised structure of the Internet opens up the freedom of movement of self-directed learning. This decentralized structure provides the infrastructural basis for associative or rhizomorphic, or hypertextual learning. The learning process arises "through the active creation of links between content-related, technical, and social resources. Accordingly, it is based on the creative social exchange about topics that the participants themselves can influence" (Grünewald et al., 2013, p. 144). This 'creative social exchange' is made possible by the collaborative potential of Web 2.0:

> But this is precisely where there is a harmonious join because connectivism, with its deliberate focus on the here-and-now reality of how digital networks support new forms of connections, social relations, and dialogue, provides a sociotechnical frame

or set of creative constraints within which contemporary social constructivist activities occur (Ravenscroft 2011, p. 144).

The 'socio-technical' descriptions of learning emphasised by Ravenscroft brings together the many-to-many structure of the Internet with learning strategies that can be understood as 'associative' with reference to Bush, as rhizomatic with reference to Deleuze and Guattari, and as hypertextual with reference to Nelson. In terms of the participatory dimension of digital cultures, self-directed learning in connectivism is thereby also production and action-oriented learning. Learning is shifted into a social context, which Ravenscroft grasps with reference to socio-constructivist approaches: It is in the social context that knowledge is constructed. Digital-based forms of collaborative learning replace the discursively established image of the individual learner sitting in front of the book in the Guttenberg Galaxy's book culture. The action- and production-oriented implications of Web 2.0-based learning in the digital age are described by Downes (2005).

In 2004, Siemens published the article "Connectivism: A Learning Theory for the Digital Age," followed in 2005 by Downes' equally powerful essay "e-Learning 2.0". Following the concept of Web 2.0, Downes formulated the concept of e-Learning 2.0. According to Downes, e-Learning 2.0 is defined by utilizing- the poly-directional and polyphonic potentials of Web 2.0 for a self-directed learning process in a social context. The innovations of Web 2.0 enable a process innovation extending to the field of learning. This process innovation has to be conceptualized in a didactically appropriate way:

> In the future, it will be more widely recognized that the learning comes not from the design of learning content but in how it is used. Most e-learning theorists are already there and are exploring how learning content-whether professionally authored or created by students can be used as the basis for learning activities rather than the conduit for learning content (Downes 2005, para. 37).

Kalz et al. (2007) emphasize that "technology for teaching and learning purposes is never didactically neutral" (Kalz et al. 2007, p. 82). With reference to Baudrillard's media theoretical considerations and the participatory or collaborative potential of Web 2.0 media, e-learning 2.0 could be understood as dialogical learning. In the course of this dialogical learning, knowledge is constructed instead of received: "In a nutshell, what was happening was that the Web was shifting from being a medium, in which information was transmitted and consumed, into being a platform, in which content was created, shared, remixed, repurposed, and passed along" (Downes 2005, para, 21).

With e-Learning 2.0, Downes describes a learning process characterized by a remix culture, which is also characteristic of a new form of collective authorship in the digital age. e-Learning 2.0 is the learning-cultural expression of a change in the leading or mass media. This change of media characterizes the transition from the electronic to the digital age. In this transformation process, the Internet is experiencing a participatory overhaul through the establishment of Web 2.0. This participatory reshaping of the Internet substitutes a receptive Internet use in the sense of Web 1.0. While the Internet was/is used in Web 1.0 according to the receptive structure of established mass media of the electronic age, Web 2.0 establishes a new form of participatory media use and knowledge construction:

> And what people were doing with the Web was not merely reading books, listening to the radio or watching TV, but having a conversation, with a vocabulary consisting not just of words but of images, video, multimedia, and whatever they could get their hands on. And this became, and looked like, and behaved like, a network (Downes 2005, para. 21, ed.).

The result of the action- and production-oriented learning culture of e-learning 2.0 is a collaborative remix culture that is medially based on the infrastructure of Web 2.0. Learning organization is supra-individual and takes the form of a network. Learning theory models such as behaviorism and, to some extent, also varieties of cognitivism, which have focused on the individual as the learning subject (see Kergel 2014), are being substituted by a collaboratively oriented understanding of learning. Learning is decentralized and collaborative. In their learning process, learners generate content such as wiki contributions, podcasts and weblogs in dialogical exchange. This content metonymically represents learning outcomes of connectivist knowledge production. Through the collective, intertextual character of these learning outcomes, the learners involved or 'connected' become part of a digitally based, collective authorship.

The decentering through the medial structuring of communication is negotiated on the levels of media theory, epistemology, and learning theory, among others. In the process, the reader or learner is enthroned and, integrated into digital-based, collaborative knowledge generation processes, becomes part of a collective authorship. Learning as a collaborative knowledge generation process presupposes the ability to engage in dialogue. Information and knowledge formation must be critically examined, modified and, if necessary, falsified in dialogue. Knowledge is generated dialogically, which leads to an opening of the text. On a textual level, this opening of the text corresponds to the ephemeral structure that characterizes the cultures of the digital and can be conceptualized with the term 'read and write culture.'

## 2.2.6   Read and Write Culture and Collective Authorship as Textual Instability

Decentralized, dialogical learning can be understood as postmodern learning, since knowledge, from an epistemological perspective, gets into 'motion' performatively or is subjected to a constant 'remix'. In this way, knowledge does not solidify in fixed (meta-)narratives or knowledge formations. Web 2.0 media offer a medial structure for such postmodern learning.

With the forms of collaborative cognition, the epistemological topos of the 'I' as a cognizing subject also erodes. The epistemological dialogue replaces the unquestionable cognitive instance of the doubting, autoreferential cogito. In dialogue, knowledge formations can be questioned from multiple perspectives. This opens up new perspectives as well as multi-perspectival possibilities for interpreting the self and the world. Petits récits can be generated in this way. The 'ideal of Wikipedia can exemplify this process of collaborative knowledge construction':

> The ideal of Wikipedia can be described as follows: Anyone who thinks a term is worth describing can generate a new entry, inviting everyone to contribute to it. The fact that several authors can write on an article simultaneously means that different knowledge is incorporated and compared with each other. All information is fed into the system and is available to everyone, due to the fact that as many experts on the subject matter as a possible write in, the most significant possible amount of information is gathered (Brandt 2009, p. 8).

Ideally, Wikipedia articles are created through collective writing; knowledge is generated and validated collaboratively. Wikipedia articles can also be understood as scientific narratives, as they arrange realities in a meaningful way and validate them intersubjectively. The fact that entries can be put up for discussion and changed at any time opens up the space of ephemeral collaborative knowledge construction. Due to its participatory approach, this form of collaborative knowledge generation defines the author/reader relation. Analogous to the implosion of time and space, Web 2.0 opens up an implosion of the author/reader relation: the recipient and the producer of content coincide. With Web 2.0, a communication structure is constituted that enables and continues Barthes' semiotic deconstruction of the author on a medial level. Weel (2011) sees in the dialogic communication structure that digital media open up the potential for a democratization of the relationship between author and reader: "Moreover, the 'democratization' of textual production, distribution and consumption creates an entirely new relationship between author and reader" (Weel 2011, p. 5). And Lessig (2009), a co-founder of the Creative Commons Initiative, also notes that a 'Read and Write Culture' has been established in the digital age specifically distinct from the 'Read Only Culture.' This Read and Write Culture constitutes a textual instability (see Weel 2011) that corresponds to the ephemeral structure of the digital cultures. "The printing press has in the course of time created a (largely unconscious) expectation of stability and permanence of form and content" (Weel 2011, p. 149). Through the participatory possibilities of Web 2.0, this text is set in motion. "Different people can comment on the same digital text, giving rise to, for example, various – virtual – combinations of texts and commentaries" (Weel 2011, p. 158). The text unfolds in a 'web' of reception and commentary. At the latest in the act of commenting, the text undergoes a remix. Each reader/author is an element of this text. A text can be received and commented on/remixed by several readers. Remix culture performatively reproduces a textual instability. With the text, the author also becomes 'unstable.'

Barthes' strengthening of the reader within the author/reader relation framework can be extended in terms of action and production theory against the background of the poly-directional and polyphonic orientation of Web 2.0: the reader himself becomes the author. He constitutes the text as a stepping stone. He reads the text and relates to it. Ideally, he comments on the text and rewrites it with the intertextual references that have inscribed themselves in the reader through the process of reception. Against the background of these considerations, the concept of the author can be perspectively substituted by the concept of collective authorship. Collective authorship refers to the possibilities of collaborative knowledge production and means a farewell to the figure of the author of the Gutenberg Galaxy's book culture. Instead of the hierarchical relation between author and reader, which is constituted by the book's unidirectional structure, the medial structure of Web 2.0 enables dialogue-based collective authorship. Such collective authorship can also be understood as postmodern authorship, as the text is successively generated through critical reflection and a dialogical relating to one another. Accordingly, Wandtke (2001) sees the technological shift towards a culture of digitality as linked to postmodernity. The media ephemeral instability of the digital corresponds to a new form of authorship that unfolds in the context of postmodernity:

> Whereas in the literary, scientific, and artistic production of industrial society, the material design (text, sound, image) is bound to a material carrier (paper, body, wood, marble, etc.), postmodernity is characterized by the computer-controlled Internet, in which by means of electronic technology new forms of material design and the communication of literary, scientific and artistic products as information in the virtual cultural space emerge (Wandtke 2001, p. 16).

The characteristics of the participatory and the ephemeral, which are significant for the digital cultures, are actualized in the textual instability and the collaborative dimension of collective authorship. It should be noted that the ideal of the collaborative form of knowledge production, which underlies the 'ideal of Wikipedia,' assumes an epistemic (postmodern) attitude on the part of the users. The ideal of Wikipedia comes close to the epistemic ideal of science. Analogous to scientific epistemological skepticism, truth claims are negotiated in Wikipedia:

> Behind [...] revisions and critiques is a universal principle of scholarship. Every discipline must be willing to question its terms; the ability to redefine terms and to prove the usefulness of the redefinition through research results is an expression of the vitality of the subject in question as a whole (Hahn 2013, p. 61 f.).

The optimism that collaborative knowledge production leads to a well-founded, epistemic knowledge construction is not shared by everyone. For example, Lanier (2010) sees the danger that collaborative knowledge production will turn the 'mob' into a subject of domination, and 'average opinions' will become conventionalized. The premise of the epistemic negotiation process of knowledge is an expression of the 'trust' (Lanier 2010, para. 3) that 'ultra-leftists have in the justice of consensus processes' (ibid.). Lanier, who among other things developed the first avatar, decisively coined the term virtual reality and received the Peace Prize of the German Book Trade in 2014, speaks in this context of a 'digital Maoism' that enthrones the mob uncontrollably:

> A collective on autopilot can be a cruel idiot, as the outbreaks of Maoist, fascist, or religious swarm minds have shown us time and again. There's no reason why such social disasters couldn't happen in the future under the guise of technological utopias. If wikis continue to gain influence, they should be improved by those mechanisms that worked quite well in the pre-Internet world (Lanier 2010, para. 14).

At this point, a discursive tension becomes evident, to which the evaluation of digitally based collaboration processes is exposed. The narrative topos of the mass as an uncontrolled, drive- or need-driven figure of movement meets a post-subjectivist concept of the digital swarm of 'swarm intelligence'. The swarm is discursively brought into the field as a form of resistance against bourgeois compulsions of individuation and forms of subjectivation. No matter how the view of collaborative forms of knowledge construction is evaluated, it can be stated that the form of authorship is at least significant in the field of discursive engagement with knowledge construction processes in the digital age. In addition to the problematization of the 'digital mass' between mob dictatorship and symbol of freedom, the

Creative Commons licenses can be read as a further indication of the media-induced shifts in the concept of the author.

In 2001, the organization Creative Commons was founded in the USA, and a license model was developed that was made publicly available. This licensing model allows the author to grant other users low-threshold usage rights to his or her work. Granting the rights of use also allows the author to explicitly agree that the content he generates can be edited and become part of new works. Creative Commons licensing makes the work free content so that free use and redistribution of the content becomes possible under copyright law. The development of free content represents an approach that originated in the Open Resource Movement, which was constituted in the course of Internet development.

> The starting point of the open-source or free software movement has initially been to program a competitive and better operating system as an alternative to commercial providers' products (first and foremost, Microsoft). Instead of the source code being written and developed by a few programmers – to the exclusion of the public – software was to be created openly and collaboratively (Brandt 2009, p. 8).

Underlying the open-source movement is the "conviction" (Stalder 2016, p. 270) "that free access to data is a necessary condition for autonomous action in the culture of digitality" (Stalder 2016, p. 270). Thus, the "practice of creating digital commons [...] goes back to the emergence of Free Software in the mid-1980s" (Stalder 2016, p. 252). Since that time, there has been a steady development of "a complex landscape that manages software codes cooperatively and sustainably as shared resources open to all – who accept the licensing terms equally" (Stalder 2016, p. 252). With reference to the open-source movement, collective authorship can be understood as part of the early Internet culture, which also manifests itself metonymically in licensing models: "Thus, to make open-source software possible at all, a special licensing system had to be developed that allowed the author to grant others further use and editing on a flat-rate basis" (Brandt 2009, p. 9). Thus, in the 1980s, the 'GNU General Public License' (GNU GPL) was developed as general permission to publish. With the GNU GPL, the idea of free software has been raised to the level of copyright licensing. With the GNU GPL, users are granted the right to reproduce and edit the software in question free of royalties and to make it publicly available again.

The Creative Commons licensing model is in the tradition of the open-source approach, and the GNU GPL model: While the open-source approach is defined by the fact that the source code of the software is freely accessible and can be further developed, Creative Commons licenses enable similar dynamics in the area of text, video and image processing. Open source stands for a range of licenses for software in common that the source code is publicly available. The requirements of Open Source software go beyond the readability/availability of the source program. The exchange of ideas for the further development of the software is an intended objective underlying the opening of the source code. The Creative Commons model continues this idea and transfers it to knowledge products/knowledge formations beyond software development. Over the years, the Creative Commons license model has been continuously developed, so that currently version 3.0 is available. This license model, in turn, has six sublicenses, which allows for a gradation of the terms of use. The gradation of the terms of use ranges from a prescribed unchangeable distribution of the contents to granting the possibility that contents may be further processed or changed. Generally, within this licensing model's framework, the author remains as the legal reference person, as he must state his name as the licensor. According to Creative Commons, this is marked with the designation CC BY, which is obligatory for every form of licensing. Creative Commons licenses can be generated with a few clicks via the Creative Commons Initiative site.[9] Under the Creative Commons licensing model, the author can prohibit his work from being edited, remixed, and then republished: CC-BY-ND (ND here stands for *Non-Derivation*). It is also possible to decree a distribution under the same conditions that one grants oneself: CC BY-SA. Here SA stands for *Share-Alike*. If, for example, a work is made available free of charge, no one else may subsequently distribute the work commercially. The individual license elements can be combined so that license models such as CC BY-NC-SA can be generated:

- Attribution to the author is required (CC BY),
- No commercial use of the content is allowed (NC) and
- The content must be redistributed in the same way as it was initially disseminated (Fig. 2.5).

**Fig. 2.5** The iconographic symbolism of Creative Commons licensing makes rights of use accessible at a low threshold, beyond written language hurdles. (Source: https://www.openaccess.uni-mainz.de/files/2015/08/by-nc-sa.eu_.png, last accessed October 12, 2017)

---

[9] See https://creativecommons.org/choose/, last accessed October 12, 2017.

With the CC0 license (Public Domain Dedication), licensors can waive all rights or grant an unconditional license. Work can then be copied, distributed in a modified form, or otherwise made available to the public. However, the legal question arises here as to whether a work may be detached from the author in this way.

The Creative Commons licensing model can be understood as an approach to reacting to the collaborative dynamics that digital-based forms of knowledge production enable. A text is no longer finitely 'fixed', but can be rewritten and distributed under certain Creative Commons license conditions. Even if the concept of the author – a phenomenon of the Gutenberg Galaxy – is not thereby abolished, it is set in motion by the possibility of editing. On the level of licensing, the Creative Commons model reflects the emerging collective authorship in the digital age, already foreshadowed semiotically in the electronic age in Barthes' deconstruction of the author. From an epistemological perspective, it can be stated that this symbolically refers to the collaborative dimension of knowledge production and the infinite movement of knowledge, or the trans-subjective dimension of knowledge generation processes.

Through its decentralized structure and dialogical orientation, the Internet provides possibilities for a collaborative knowledge generation process that can lead to collaborative authorship. The epistemological shifts caused by the media structure of the Internet can also be seen in the thematization of the Internet as a space of freedom. The deconstruction of the subject as an autonomous instance of knowledge, which took place in post-structuralism, finds here a media-theoretically refracted actualization. This actualization of the deconstruction of the subject became increasingly popular in the context of the discussion about cyberspace as a space of freedom in the 1990s and continues to shape the protest cultures and forms of resistance that are emerging on the Internet today. This aspect will be discussed in the following.

## 2.3    Cyberworld as a Space of Freedom – 'The Assignable Self Becomes Fluid'

In the sense of a totality-oriented concept of culture, culture is characterized by how individuals relate to the world (how they appropriate the world, deal with it, and thus change it or bring it forth [hervorbringen]). With digitalization, this world is changing, and with it, the cultural practices. Therefore, to look at the relationship between the individual and digitalization from a cultural theory perspective, a media theory-oriented analysis of the epistemological relationship between the individual and the Internet is relevant:

Is the Internet a 'cyberspace' and thus an alternative world into which the individual enters through digital media, or is it impossible to separate a material-physical world from a 'virtual world'?

At the latest, with the introduction of the first graphics-capable web browsers at the beginning of the 1990s, the Internet is increasingly becoming part of reality. The Internet's examination leads to the question of the significance of this new media dimension of reality – and how the Internet generates cultural meaning.

Reichert (2013) notes that with the "development of web browsers for graphical user interfaces [...] beginning in the mid-1990s, the Internet had become graphic" (Reichert 2013, p. 29). With the emergence of graphical browsers, "emancipatory discourses on networking culture emerged" (ibid.). These discourses on the Internet's cultural significance are thereby tied to the development of the Internet's media structure: from the beginning of the 1990s until the mid-2000s, the Internet was primarily tied to the desktop PC. Sitting in front of the desktop PC, it was possible to dial into the world of the Internet. *Cyberspace marks this separation between the material-physical world and the virtual world of the Internet, bound to technology.*

The term cyberspace comes from the science fiction author William Gibson, who used the word Cyberspace in the 1984 novel 'Neuromancer.' Cyberspace is a combination of the words' cybernetics' and 'space.' The concept of Cyberspace is open to interpretation and offers a projection surface for utopian occupations. Against the background of this openness to interpretation, an ultimate conceptual framing of Cyberspace does not seem possible. On a conceptual level, the ephemeral dimension that constitutes digital culture manifests in this way:

> Cyberspace is above all a diffuse place, categorically weakly defined, ontologically fuzzy, and still highly modest in its phenomenological offerings for those hungry for experience even beyond its technically fragile construction. This creates the best conditions for populating this strange place with countless human fantasies that celebrate their resurrection or revival here (Palm 2006, p. 59).

Since the mid-1990s, Cyberspace has increasingly been discursively discussed as a space of freedom; the separation of the physical-material world and Cyberspace is semiotically coded. Hartmann (2000) suggests that the "'lawless' electronic space with its unrestricted freedoms [...] together with the evocation of new commonalities probably represents a phantasm of the pioneering American spirit" (Hartmann 2000, p. 318). Discursively, the virtual world of the Internet represents a counter-world to the normative constraints of the material-physical world. Paradigmatically, this discursive constellation was performed by Barlow with "A Declaration of the Independence of Cyberspace," published on February 8, 1996. The "Declaration of the Independence of Cyberspace" criticizes state access to the Internet, which is discursively staged as a post-national and post-state space of freedom. The concrete occasion for the publication of the Declaration of the Independence of Cyberspace was the US government's Telecommunications Act in 1996. In addition to deregulation decisions in the area of radio and television and in the area of mobile and fixed network companies, the law contained an amendment, the so-called 'Communication Decency Act,' which referred to content published on the Internet. This amendment was intended to make it possible to regulate and control pornography on the Internet more strongly by, among other things, prohibiting the posting of indecent and 'obviously offensive' content. Critics have argued that this would lead to Internet censorship. The "Declaration of the Independence of Cyberspace" was published by Barlow as co-founder and representative of the Electronic Frontier Foundation (EFF). As a representative of the EFF, Barlow acted as a representative of an NGO that advocates human rights in the Internet age. In the context of the "Declaration of the Independence of Cyberspace," Cyberspace is staged as a counter-world to the material-physical world's established power structures. The Declaration of Independence opens with the words: "Governments of the Industrial World, you weary giants of flesh and steel, I come from Cyberspace, the new home of Mind. On behalf of the future, I ask you of the past to leave us alone. You are not welcome among us. You have no sovereignty where we gather" (Barlow 1996, para. 1). The post-nationalist space of freedom in Cyberspace is influenced by traditional anarchist narrative topoi, which are based on a dichotomous opposition between state and society[10]:

> We have no elected government, nor are we likely to have one, so I address you with no greater authority than that with which liberty itself always speaks. I declare the global social space we are building to be naturally independent of the tyrannies you seek to impose on us. You have no moral right to rule us, nor do you possess any methods of enforcement we have true reason to fear (Barlow 1996, para. 2).

Instead of laws, Cyberspace is characterized by its 'own' culture and ethics, which are reproduced performatively through interactions (Barlow 1996, para. 4). Barlow locates this culture or world of Cyberspace beyond "race, economic power, military force, or station of birth" (Barlow 1996, para. 7). Established forms of "legal concepts of property, expression, identity, movement, and context do not apply to us. They are all based on matter, and there is no matter here" (Barlow 1996, para. 9). Via this dichotomizing form of argumentation, Cyberspace is positioned as a

---

[10] A paradigmatic example could be the main theoretical work "The Liberation of Society from the State" (1932) by the anarchist Erich Mühsam.

counter-space to the material-physical world and semiotically encoded as a space of freedom. In this, the forms of interaction and organization of Cyberspace are "a standing wave in the web of our communications" (Barlow 1996, para. 6). Through the implosion of space, actors in Cyberspace escape the spatial fixations they face as individuals in the material-physical world: "Ours is a world that is both everywhere and nowhere, but it is not where bodies live" (ibid.). As a space of freedom, Cyberspace is not accessible to governmental practices.

While Barlow discusses the dichotomous setting between Cyberspace and the material-physical world to establish a post-nationalist space of freedom, social psychologist Turkle discusses this dichotomous setting in the field of identity construction. In her work "Life on the Screen: Identity in the Age of the Internet" published in 1995, Turkle states: "We can easily move through multiple identities, and we can embrace – or be trapped by – Cyberspace as a way of life (Turkle 2011, p. 231). Turkle sees the Cyberspace of the Internet as providing a 'play space.' Individuals can break free from their social roles and the normative constraints that accompany the social roles, metonymically manifested in role expectations. As a paradigmatic example of the ephemeral possibilities of Cyberspace, Turkle refers to so-called multi-user computer games or MUDs. MUDs are a transfer of text adventures into the digital world. The first MUD was programmed in 1978 in Essex by the student Roy Trubshaw. It was based on the role-playing game 'Dungeon,' which was initially designed for only one player (this also explains the original abbreviation MUD, which stands for '*Multi*-User Dungeon'). MUDs are text-based role-playing games where players dial into a game server. In order for characters to interact in the virtual world of the MUD, players enter commands via keyboard to which the software responds: "MUDs put you in virtual spaces in which you can navigate, converse, and build" (Turkle 2011, p. 11). Turkle anticipates the typical Web 2.0 intersection of content recipients and consumers concerning the requirements of MUD gamers:

> MUDs are new kind of virtual parlor game and a new form of community. In addition, text-based MUDs are a new form of collaboratively written literature. MUD players are MUD authors, the creators as well as consumers of media content. In this, participating in a MUD has much in common with script writing, performance art, street theater, improvisational theater (Turkle 2011, p. 11 f.).

Turkle sees the subversive potential of MUDs in escaping the individuation constraints of bourgeois society – "To each individual his place and to each place an individual" (Foucault 1997, p. 183):

Today I use the personal computer and modem on my desk to access MUDs. Anonymously, I travel their rooms and public spaces [...] I create several characters, some not of my biological gender, who are able to have social and sexual encounters with other characters. On different MUDs, I have different routines, different friends, different names (Turkle 2011, p. 15).

The individual – constituted in bourgeois society as a monad of social hierarchies – is dissolved: Cyberspace is interpreted as a space of freedom that enables multiple and fluid identity constructions. To analyze epistemologically these new possibilities of identity construction in Cyberspace, Turkle draws on post-structuralist approaches:

Thus, more than twenty years after meeting the ideas of Lacan, Foucault, Deleuze, and Guattari, I am meeting them again in my new life on the screen. But this time, the Gallic abstractions are more concrete. In my computer-mediated worlds, the self is multiple, fluid and constituted in interaction with machine connections (Turkle 2011, p. 15).

Turkle uses post-structuralist approaches to conceptualize Cyberspace as a space of freedom – whereby post-structuralist approaches are "often mentioned in the same breath" (Angermüller 2013, p. 247) as a "postmodernist cultural" (ibid.) approach. In this context, the 'ephemeral programmatic' of Deleuzian epistemology (including the dissolution of the individual in collaborative contexts) corresponds to the ephemeral self-experience that, according to Turkle, MUDs make possible: "MUDs make possible the creation of identity so fluid and multiple that it strains the limits of the notion Identity, after all, refers to the sameness between two qualities, in this case between a person and his or her persona. But in MUDs, one can be many" (Turkle 2011, p. 12). The possibility of multiple identity designs can be realized through anonymity in MUDs. In this way, anonymity in Cyberspace makes it possible to escape the individuation constraints of the physical world: "The anonymity of MUDs – one is known on the MUD only by the name of one's character or characters – gives people the chance to express multiple and often unexplored aspects of the self, to play with their identity and to try out new ones" (Turkle 2011, p. 12).

When Turkle resorts to the metaphor of the 'liquid' to describe the forms of identity construction processes in 1990s Cyberspace, she is updating a metaphor used in the field of epistemological reflection by Deleuze. Deleuze uses 'becoming liquid' to describe the dissolution of the individual as the social vanishing point of subjectivizing practices:

Individuation is no longer enclosed in a word Singularity is no longer enclosed in an individual […] You see, the forces of repression always need a Self that can be assigned, they need determinate individuals on which to exercise their power. When we become the least bit fluid, when we slip away from the assignable Self, when there is no longer any person on whom God can exercise his power or by whom He can be replaced, the police lose it. This is not theory (Deleuze 2004, p. 138).

The fluid is contrasted with the assignation of the individual in the material-physical world. Unlike the discourse on the counter-public, Cyberspace as a space of freedom means a step beyond protest: in the most real sense of the word, Cyberspace serves as utopian space/a space without place. Among the "hackers and pioneers of computer networks, the notion of communication media as a space for action" emerges (Stalder 2016, p. 79). This notion is in the tradition of the freedom movements of the 1960s and 1970s, in which "alternative forms of life and organization" (ibid.) were sought. Standing in this line of tradition, Cyberspace is discursively thematized as a space of freedom. An understanding/a concept of counterculture is continued that was formulated in the 1960s and the 1970s. The genesis of hacker culture can exemplify this.

The origins of the term hacker can be traced to student subcultural practices at the Massachusetts Institute for Technology (MIT). Originally used since the mid-1950s to refer to the innovative use of equipment in the field of amateur radio, hacking was used by MIT's Tech Model Railroad Club in the late 1950s. The term hacker/hacking referred to an innovative use of electronic and mechanical equipment. In the discussion about computers, the innovative, sometimes misappropriated handling of computers was called hacking. The word hack – also in the context of student subculture – acquired a subversive connotation and denoted mostly technology-based pranks. A collaborative dimension characterizes the spirit of innovation that signified the use of technical devices and later software under the sign of hacking. Particularly in the context of a hacker culture located at universities, source codes were exchanged – in software development, for example. In this context, software development was a collaborative process of knowledge generation and the ground from which the open-source movement developed: Richard Stallman, formerly of MIT's AI Lab, founded the GNU Project in 1983 and was involved in developing the GNU GPL, the open-source movement's legal foundation. In 1985, Stallman co-founded the Free Software Foundation (FSF), which provided an institutional framework for the GNU Project.

The collaborative spirit of hacker culture led to a digital-based understanding of knowledge construction, which also manifests itself in the Creative Commons approach. The hacker culture organized itself, among other things, in the form of 'clubs' such as the Homebrew Computer Club (HCC), founded in 1975. Among others, Steve Wozniak belonged, who founded the company Appel with Steve Jobs. 'Computer lovers' and 'technology enthusiasts' met within the framework of these hacker clubs. The best known German hacker club is the Chaos Computer Club (CCC). It was officially founded in 1981 and emerged from the Sponti movement: The Sponti movement's main event is the 'Tunix'-Congress (Do-Nothing-Congress). The Tunix-Congress was a congress of alternative culture, which took place with about 15,000 participants from January 27 to January 29, 1978, at the Technical University in Berlin. The formulation of the congress as "Treffen in Tunix" (Meeting in Doing Nothing) refers to subversive irony, which can be deciphered as a form of postmodern protest and characterizes the spontaneous slogans. From the congress, which was part of the aftermath of the West German protest movements of the 1960s and 1970s, important impulses emanated on the unfolding alternative culture of the 1980's. The left-alternative newspaper project 'Taz' was introduced, and the idea of a nationwide ecology party, from which 'the Greens' later emerged. In 1981, the 'Tuwat Congress' (Do Something) took place. The name was an ironic reference to the meeting in Tunix. The congress opposed the eviction of houses announced by the Berlin Senate. The CCC was founded during this congress. In the self-description of the founding process, explicit reference is made to 'insignia' of alternative cultures such as the 'desk of commune 1' and the 'Taz':

> The roots of the Chaos Computer Club go back to 1981. Back then, hackers, self-described "Computerfreaks," met at the table of Kommune 1 of the newspaper Taz in Berlin as part of the "tu-wat" congress. They had recognized the emerging electronic data networks' possibilities and wanted to put them to creative use and exchange ideas. The meeting initially led to the Chaos Computer Club's informal founding and was continued as a regular meeting in Hamburg (Chaos Computer Club n.d., para. 1).

The CCC, anchored in the roots of the alternative culture that emerged in the aftermath of the 1960s and 1970s, gradually developed into an influential association of hacker culture. The CCC, of which Edward Snowden is an honorary member, organizes congresses, issues publications, publishes podcasts, etc. In the form of symbolically effective hacks, the CCC campaigns for data protection have already written expert opinions for the German Federal Constitutional Court. Representatives of the CCC have been invited as experts to hearings of the federal government.

The CCC shows how hacker culture is becoming increasingly politically involved in social issues. This explicitly political orientation of hacker culture is also reflected in hacktivism, an amalgamation of the terms hacker and activism. Hacktivism involves the targeted use of hacking strategies and techniques for political ends – such as when messages are left on highly symbolic, highly trafficked websites. Another characteristic that distinguishes hacker culture is the play with synonyms. In this way, a free space is created in Cyberspace, making it possible to distance oneself from one's own civic identity, which is metonymically fixed via first and last names. In terms of discourse analysis, this treatment of anonymity can be demonstrated by the relational structure between a book about the future of the Internet and this book's discussion. An interview conducted in the course of gathering material for the book in question: Cypherpunk, hacker, and founder of Wikileaks, Julian Assange, is interviewed about anonymity on the net in an interview that included Eric Schmidt (the former Executive Chairman and Chief Executive Officer of Google). This interview was used to collect data for the book "The New Digital Age: Reshaping the Future of People, Nations, and Business" (2013), which Schmidt published with Jared Cohen, the Director of Google Ideas. One of the issues addressed in this book is that of data security in the digital age. The position taken is that while access to data is a threat to privacy, it is not threatened in democratic states. Democratic regulatory processes would correct the transgression of privacy: "Governments operating surveillance platforms will surely violate restrictions placed on them [...] eventually, but in democratic states with properly functioning legal systems and active civil societies, those errors will be corrected whether that means penalties for perpetrators or new safeguards put into place" (Schmidt and Cohen 2013, p. 175). Assange criticized the book in a book review published in the New York Times as euphemizing a colonizing, capitalist practice of power. In doing so, Assange points to an entanglement between state control and the control of large corporations like Google to which individuals are subjected. In a condensed way, Assange describes in the book review a power practice of the control society in the digital age:

> The authors offer an expertly banalized version of tomorrow's world: the gadgetry of decades hence is predicted to be much like what we have right now – only cooler. "Progress" is driven by the inexorable spread of American consumer technology over the surface of the earth. Already, every day, another million or so Google-run mobile devices are activated. Google will interpose itself, and hence the United States government, between the communications of every human being not in China (naughty China). Commodities just become more marvelous; young, urban professionals sleep, work and shop with greater ease and comfort; democracy is insidiously subverted by technologies of surveillance, and control is enthusiastically rebranded as "participation"; and our present world order of systematized domination, intimidation and oppression continues, unmentioned, unafflicted or faintly perturbed (Assange 2013, para. 13).

Assange's resistance is to these practices of control and the ideologization of surveillance. Anonymity enables a resistance strategy. In doing so, Assange updates
the narrative topos of anonymous freedom of movement in Cyberspace, which
positions itself beyond the disciplining subjectification constraints of the material-
physical world. According to Assange, resistance is characterized by a guerrilla-
like ephemeral movement strategy. Assange describes this movement strategy in
the interview with Cohen and Schmidt, which they conducted with Assange in
preparation for their book "The New Digital Age: Reshaping the Future of People,
Nations, and Business":

> I am always hesitant in saying that everyone should go out and become a martyr. I
> don't believe that. I believe the most effective activists are those that fight and run
> away to fight another day, not those who fight and martyr themselves. That's about
> judgement — when to engage in the fight and when to withdraw so as to preserve
> your resources for the next fight (Assange 2014, p. 137).

This ephemeral freedom of movement is made possible by anonymity – which
hackers, for example, discursively represent through synonyms: "If you have perfect anonymity, you can fight forever, yes. You don't have to run away" (ibid.). As
a strategy of resistance, which Assange describes here, this anonymity was aes-
thetically thematized by cyberactivists from Anonymous and consistently elevated
to an organizational principle.

### 2.3.1  We Are Many! – Anonymous Collectivity

Against the background of post-structuralist difference theory and with reference
to Turkle, Cyberspace can be conceptualized as a space that enables new 'modes of
life.' New 'modes of life' in turn "inspire ways of thinking; modes of thinking create ways of living" (Deleuze 2004, p. 66). Discursively, an understanding of
Cyberspace is created here that defines Cyberspace as a space beyond the constraints of subjectification. In the course of the discursive thematization of
Cyberspace, traditional topoi of freedom are updated. In the process, the legacy of
the protest movements of the 1960s and 1970s continues to have an effect: "With
the emergence of the Internet, the digital medium becomes the new carrier of networking utopias, in which the unfulfilled social hopes of alternative cultures are
perpetuated, especially from the mid-1990s onwards" (Friedrich and Biermann
2016, p. 83).
    Münker (2009) points out that "[t]o the net utopians of the nineties [...] the
Internet appeared as a medium for establishing a radically different world" (Münker

2009, p. 66). One feature of the freedom potential of the Cyberspace of the Internet represents the detachment from subjectification constraints through anonymity. The potential anonymity of users is seen as positive from an emancipatory, power-critical approach. Although the thesis of Cyberspace as a subversive counter-world was formulated in the mid-1990s, this discourse developed an impact that continues into the present. The topicality of anonymous resistance can be seen in the emblematic self-description "We are Anonymous":

> We are legion/many
> We do not forgive
> We do not forget
> Expect us

In this discursive thematization, often played at the end of Anonymous messages, the release of individuation constraints is formally accomplished by using the first person *plural*. The semantics of anonymity through the first person plural is supported by Guy Fawkes masks worn by Anonymous members or depicted on Anonymous messages. Rather than having an identifiable face, the individual disappears into the mass of the plural of the Guy Fawkes mask – which in turn is part of pop cultural iconography: as a historical figure, Guy Fawkes was a Catholic officer who attempted to carry out an explosive assassination on King James I and the English Parliament in London on November 5, 1605. Through the comic book "V for Vendetta" ("V for Vendetta"), the character Guy Fawkes became part of popular culture, and the Guy Fawkes mask became a pop-cultural icon of resistance. "V for Vendetta" originated as a comic narrative during the 1980s in England, which was increasingly dominated by neoliberal and authoritarian social policies due to the Thatcher government (Fig. 2.6).

The narrative is set in a fictional 1997 England, where the anonymous and anarchist lone wolf V wages a guerrilla-like war against a totalitarian state. This state has been constituted by a fascist party's takeover of England after World War III. The anarchist resistance fighter V represents the right to self-determination and is a counter-design to hegemonic powers associated with foreign domination. In 2005, this comic narrative was made into a film by the Wachowski siblings. V increasingly became part of a left-oriented, non-dogmatic pop cultural iconography that the Anonymous movement took up:

One can well understand the reasons why *Anonymous* 'chose' this character for identification: 'V' simply represents a perfect embodiment of the representatives of cyberculture: He fights against a centralized government, he is a determined, independent

**Fig. 2.6**  The Guy Fawkes
mask becomes a symbol of
anonymous collectivity.
Through the pop-cultural
references, a spontaneist form
of resistance is discursively
actualized. (Source: https://i.
pinimg.com/736x/94/24/ce/9
424ce215c038bb46dbbcc8cfd
003a09%2D%2Danonymous
-mask-gy-fawkes.jpg, last
accessed: October 12, 2017)

*tech-savvy* – misunderstood in a society closed to political ideas that he seeks to defend for the good of all (Bardeau and Danet 2012, p. 107, ed.).

V, who always wears a Guy Fawkes mask, becomes the pop-cultural projection
surface of an "'Anonymous Collectivity' on the Social Net" (Reichert 2013, p. 11).
Through such an Anonymous Collectivity, a form of subversive resistance is established that "strictly rejects identity and representation" (ibid.). This form of resistance has tradition and was prepared by the hacker movement. For example, in
1986, 10 years before the declaration of independence of the Internet, the hacker
'The Mentor' alias Loyd Blankenship formulated the "Hackermanifesto – The
Conscience of a Hacker." The hacker manifesto was published in the underground
magazine 'Phrack' and features significant topoi that would shape Cyberspace's
discourse as a libertarian space in the 1990s. In the hacker manifesto, the Internet
is staged as a power-free counter-world to the material-physical world:

> And then it happened... a door opened to a world... rushing through
>     the phone line like heroin through an addict's veins, an electronic pulse is
> sent out, a refuge from the day-to-day incompetencies is sought... a board is
> found.
> "This is it... this is where I belong..."
> I know everyone here... even if I've never met them, never talked to
> them, may never hear from them again... I know you all... (The Mentor 2004, para. 9).

The description of the experience of the counter-world ties in with the semantics of
the discourses around drug experimentation that emerged in the course of the protest culture of the 1960s and 1970s. As a home, the new world offers a new form of
a community located beyond everyday powerlessness. This new community dis-

tances itself from capitalist profit maximization: "We make use of a service already existing without paying for what could be dirt-cheap if it wasn't run by profiteering gluttons, and you call us criminals" (The Mentor 2004, para. 12). The dichotomization between the Internet and the material-physical world is metonymically tied to the negotiation of the concept of the criminal:

> We explore... and you call us criminals. We seek after knowledge... and you call us criminals. We exist without skin color, without nationality, without religious bias... and you call us criminals.
> You build atomic bombs, you wage wars, you murder, cheat, and lie to us and try to make us believe it's for our own good, yet we're the criminals.
>
> Yes, I am a criminal. My crime is that of curiosity. My crime is that of judging people by what they say and think, not what they look like.
> My crime is that of outsmarting you, something that you will never forgive me for. (The Mentor 2004, para. 12f.).

A criminal is someone who, from a legal perspective, is guilty of a crime. In the argumentation of The Mentor, two forms of criminality collide: the legally coded hierarchy and dependency relationships – in other words, the power relations of the material-physical world are in tension with the libertarian ideals made possible and embodied by Cyberspace.

With the hacker manifesto, the battle for the Internet's symbolic order has begun: Who or what behavior is considered criminal on the Internet? This is a central question relevant to the discussion of copyright protection in the digital space of Cyberspace, which prompted the hacktivists of Anonymous, among others, to launch 'Operation Payback' in 2010–25 years after The Mentor's hacker manifesto (Fig. 2.7).

The Mentor sees the ideals of freedom of the Internet beyond individuation characteristics or constraints that lead to a disciplinary focusing of the individual in the sense of Foucault ("We exist without skin color, without nationality, without religious bias ... and you call us criminals", The Mentor 1986, para. 12). Accordingly, the claim to freedom formulated by the hacker manifesto cannot be suppressed by acting against its author. Thus, the manifesto concludes with the statement: "I am a hacker, and this is my manifesto. You may stop this individual, but you can't stop us all.... after all, we're all alike" (The Mentor 2004, para. 14). Bardeau and Danet (2012) see the manifesto as a "connecting line to *Anonymous*" (Bardeau and Danet 2012, p. 15, e.i.o.). The decentralized structure of the Internet allows a non-taxonomic, rhizomorphic communication and, to escape the individuation constraints of the material-physical world. Castells (2004) points out that this

**Fig. 2.7**  Anonymous stages the detachment from individual identities by resorting to patterns of iconic pop culture. The motto refers to collective resistance through the reference to the anonymous 'we' that shows solidarity. The image of the question mark in the place where a face would stand underlines the detachment from concepts of individuation in the symbolism of the visual language. (Source: http://pcwallart.com/anonymous-logo-transparent-wallpaper-2.html, last accessed October 12, 2017)

form of communication was structurally inherent in the Internet and was even intended by the initiators – the US military:

> As is well known, the Internet traces its origins to a daring plan hatched in the 1960s by the technological warriors of the US Defense Department Advanced Research Projects Agency (the mythical DARPA) to prevent the Soviets from taking over or destroying American communications systems in the event of nuclear war. To some extent, this was the electronic counterpart of the Maoist tactic of dispersing guerrilla forces over a vast territory to counter enemy power with flexibility and terrain knowledge. The result was a network architecture that, as intended by its inventors, could not be controlled by any center and consisted of thousands of autonomous computer networks that had innumerable ways of linking up and circumnavigating electronic barriers (Castells 2004, p. 6 f.).

To describe the organizational structure of the Internet, Castells resorts to the organizational form of 'guerrilla forces.' This describes the "socius structure" (Fischer 2006, p. 22) of the Internet, i.e., the forms of communication and action made possible by the Internet: Due to its decentralized structure, the Internet opens up spaces for a guerrilla-like resistance against hegemonic powers in Assange's sense.

Guerrilla warfare can be defined as a form of resistance in which a small group of fighters wages a decentralized struggle against organized forces. Instead of uniforming themselves, guerrilla fighters often disguise themselves as civilians. They attack unexpectedly and suddenly and retreat again. This form of combat makes it difficult for soldiers in an organized army to successfully engage guerrilla fighters in a sustained way. In the 1960s and 1970s, guerrilla warfare was discursively increasingly staged as a form of armed left-wing resistance against hegemonic power. Thus, guerrilla warfare, and as its iconic embodiment in Che Guevara, became a symbol of anti-hegemonic resistance.

The decentralized structure offers a 'cyber-guerrilla' a suitable media infrastructure. Accordingly, Reichert points to a content-form dynamic between the manner of medial organization and the constitution of resistance through an Anonymous Collectivity. Reichert states that "[i]n analogy to the openness and dynamics of distributed networks [...] a figure of thought of collectivity was designed" (Reichert 2013, p. 13), "which constitutively eludes identification and representation" (ibid.). This Anonymous Collectivity of cyber-guerrilla manifests itself in the hacktivist movement 'Anonymous.'

Anonymous emerged from the chat website '4chan'. 4chan is characterized by the fact that pictures can be posted openly, some of which have bizarre, grotesque, or provocative content. The threads, i.e., a series of discussion posts, are deleted relatively quickly (sometimes after hours). 4chan favors ephemeral non-hierarchical communication, in which the authors of the posts can also remain anonymous. Only the logfiles and IP addresses are stored as user data as long as the posts exist. In the communication in 4chan, authors can post contributions without a registration procedure. Posts of unregistered users are marked with the name 'Anonymous.' Through this procedure, "[h]orizontal forms of an organization [...] emerged, based on anonymity in the threads and on a staunch defense of free expression" (Bardeau and Danet 2012, p. 71). This is probably where the name of Anonymous originates: the first hacktivists of Anonymous came from the environment of 4chan or have been users of 4chan and directed their first hacking attack against a social network in 2006: the so-called 'Habbo Raid' represents the first action attributed to Anonymous. On July 12, 2006, the social network Habbo, founded in 2000, was attacked. As online community for teenagers, Habbo integrates chat and online gaming features. Users can create avatars – called Habbos – and set themselves up in a virtual hotel. A rumor arose that the online community's moderators were taking more decisive action against dark-skinned avatars/Habbos. As part of the attack on the network, access to the virtual pool was blocked. Such actions can be interpreted as the first manifestations of the constitutional process of Anonymous. However, at that time, i.e., between 2006 and 2007, "hacktivism, if one can speak of such at *4chan, was* still uncertain and not very self-confident" (Bardeau and Danet 2012, p. 71, e.i.o.). Bardeau and Danet (2012) point out that "until 2008 or

2009 [...] *Anonymous* [was] not united by a clear and structured discourse that would allow them to be given substance" (ibid, e.i.o.). Thus, one can rather speak of 'selective topics' and a 'mindset,' within the framework of which Anonymous gradually emerged from 4chan (ibid.).

The first media-effective form of Anonymous' Internet protest is the project 'Chanology.' Chanology, an amalgamation of the words 4chan and Scientology, marks the decisive step that allowed Anonymous to become an internet-based protest and resistance phenomenon: "Project Chanology was a pivotal point in the history of Anonymous in the sense that it marked its activist turn" (Firer-Blaess 2016, p. 27). The starting point was a video featuring Tom Cruise: in January 2008, a video not intended for the public was uploaded to YouTube. In the video, Tom Cruise made uncritical comments about Scientology. Scientology demanded that the video be deleted, arguing that it infringed copyright. In response to this demand, Anonymous proclaimed 'Project Chanology,' which was committed to resisting Internet censorship. Among other things, Scientology websites were attacked through Denial of Service attacks (DDos)[11] – among others, the deliberate overloading of a server through an overload of traffic. Scientology reacted by hiring the security firm Prolexic, whereupon the protests shifted. There were public, international demonstrations in front of Scientology headquarters in January to March 2008. To protect their anonymity, demonstrators masked themselves with Guy Fawkes masks, among other things, and transported the topos of the anonymity of Cyberspace into the streets. In a press release by Anonymous, the war was declared on Scientology. This was justified with the protection of freedom of speech and the objective of combating the financial exploitation of Scientology members.

Through the Chanology action, Anonymous took on a "recognizable form" (Bardeau and Danet 2012, p. 71). The protest against Scientology "produced images and symbols to which the Anonymous activists attached themselves. At the same time, they became identifiable to the general public and the media, which helped to codify the movement's codes" (ibid.). Following Chanology, Anonymous carried out further acts of cyberprotest. Since 2010 the so-called 'Operation Payback' has been running, in the course of which DDos attacks were carried out

---

[11] Distributed Denial of Service attacks are among the most common cyberattacks. The goal of these attacks is a deliberately induced overload of the IT infrastructure, as a result of which a requested service is no longer available or only available to a very limited extent.

on websites of the Motion Picture Association of America (MPAA),[12] the Recording Industry Association of America (RIAA),[13] and the company Airplex Software, which offers services against 'net piracy'. The starting point was the Indian cinema industry's order to Airplex Software to take action against websites that allowed films to be downloaded in violation of copyright. Operation Payback expanded from December 2010 into a solidarity action for Wikileaks and became 'Operation Avenge Assange.' This action represented Anonymous reaction to the so-called 'account freeze' against the whistleblower platform Wikileaks: In November 2010, Wikileaks began publishing over 200,000 dispatches from US embassies. A political echo followed these publications: "The FBI, followed by the army and the US judiciary, brought lawsuits against *WikiLeaks,* accusing it of espionage and threatening the security of the United States" (Bardeau and Danet 2012, pp. 89 f., ed.). Another reaction to the disclosures was that several companies refused to continue executing financial transactions related to Wikileaks. For example, Bank of America stated that Wikileaks might be involved in activities that were not in accordance with Bank of America's corporate policies. As a result, Bank of America refrained from executing wire transfers to Wikileaks in the future. Swiss bank Post Finance and credit card companies Visa and Mastercard reacted similarly. And the online payment service PayPal also refused to accept donations for the Wau Holland Foundation, which passes funds to Wikileaks. Since Wikileaks, as a noncommercial organization, is funded by donations, such a refusal means massive difficulties in obtaining funds. In response to this account freeze, Anonymous announced on December 27, 2010, that it would launch a DDos attack on Bank of America. As a result, the bank's main website was irregularly accessible. Similar attacks had already taken place on the websites of Visa and Mastercard on December 8 – and were critically commented on via Twitter by Barlow, who published the Declaration of the Independence of the Cyberspace: "Sorry, but I don't support DDoSing Mastercard.com. You can't defend The Right to Know by shutting someone up".[14] In addition to the shutdown of financial services, Wikileaks' technical infrastructure was also attacked. On December 2, 2010, the IT service

---

[12] The MPAA is a central association of the US film industry. Among others, Paramount Pictures, Warner Bros, Sony Pictures Entertainment, Walt Disney Motion Pictures Group, Universal Studios and twentieth Century Fox are organized within the framework of the association.

[13] The RIAA is an influential association or community of interest representing companies in the US music industry. Among other things, the RIAA awards prizes for record sales.

[14] See https://twitter.com/jpbarlow/status/12521072775663616, last accessed August 6, 2017.

provider Every DNS had discontinued its service for Wikileaks. Every DNS until then ensured the connection between the Wikileaks web presence and Wikileaks' servers. Amazon refused to provide storage space to Wikileaks, leading to the tweet from Wikileaks – "If Amazon is so uncomfortable with the first amendment, they should get out of the business of selling books."[15] In response to the undermining of Wikileaks' technical infrastructure, other websites offered storage space to Wikileaks, while Amazon was also subjected to DDos attacks. In the course of this confrontation, a dichotomous opposition of forces is discursively established, which confront each other in the fight for (freedom) spaces of the Internet. Bardeau and Danet exemplify such a dichotomous representation of this conflict:

> On one side, you find the governments, Western democratic or more authoritarian, and the multinationals. On the other side, *Wikileaks,* the supporters of the *open-data* movement and the NGOs fighting for more transparency and more freedom of expression position themselves. This scheme may seem simplified, but it concretely outlines the existing forces (Bardeau and Danet 2012, p. 93, e.i.o.).

From this perspective, Anonymous metonymically formulates a radical claim to freedom that enters into a struggle for the semiotic coding or the symbolic order of the Internet against the 'powers' of the material-physical world.

## 2.3.2  The Organless Body of the Digital Swarm

The concept of symbolic order as an analytical category describes the structuring of meaning in society. Symbolic order conceptually captures the discursive manifestation/legitimation of social hierarchies or dependency relations:

> The use of symbols – words, gestures, objects – is so powerful in creating power primarily because they stand in place of something else that does not need to be made explicit. These symbols are recognized and acknowledged as legitimate by all participants. Consequently, the basis of the production of such a symbolic, classificatory order is a certain tacit agreement and thus a kind of complicity of the dominated (Suderland 2014, p. 127).

Accordingly, Zizek (2011), with reference to Lacan, understands the symbolic order as the "unwritten constitution of society" (Zizek 2011, p.18) in which every individual is embedded. It "is here, guiding and controlling my hand; it is the sea

---

[15] See  https://twitter.com/wikileaks/status/10073870316863488?lang=dezuletzt,  accessed August 9, 2017.

in which I swim" (ibid.). In the symbolic order, social hierarchies manifest them-selves, and the logic of relations of domination are formulated. Hierarchies and relations of dependency are discursively produced or legitimated in the symbolic order. From this perspective, the symbolic order describes an arena of power. Within the semiotic system spanned by a symbolic order, space also opens up for resistance or, in the sense of the narrative logic of postmodern epistemology, space for a protesting' counter-speech.' The symbolic order is consequently character-ized by a dynamic between

- performative (re)production of hierarchies and dependency relations on the one hand and the
- articulation of resistance in the space of the symbolic order on the other hand

The concept of symbolic order can also be analytically applied to Anonymous and its fight against Internet censorship. The constellation of the conflict or struggle in which Anonymous and Wikileaks are located can be analyzed: The freedom poten-tials of the Internet enter into conflict with the hierarchies and dependency struc-tures or power structures of the material-physical world. Assange et al. (2012), member and former spokesperson of the Computer Chaos Club, notes in discus-sion with Assange and the Wikileaks activists at the time, Jacob Appelbaum:

> If you look at the internet from the perspective of people in power then the last twenty years have been frightening. They see the internet like an illness that affects their abil-ity to define reality, to define what is going on, which is then used to define what the people know of what is going on and their ability to interact with it (Müller-Maghun in Assange et al. 2012, p. 23).

Regarding the dichotomous juxtaposition between the freedom potential of Cyberspace on the one hand and the hegemonic powers of the material-physical world (such as states and large corporations) on the other, Anonymous can be in-terpreted as an actor in the 'struggle' for the semiotic encoding of the Internet. From this perspective, Anonymous continues the legacy of the freedom semantics of the 'net utopians' of the 1990s (see Münker 2009, p. 66). Following Barlow's Declaration of Independence of the Internet and the hacker manifesto of The Mentor, the freedom space of Cyberspace is defended by Anonymous against en-croachments by power holders of the material-physical world. This struggle for the Internet's symbolic order is exemplified in Operation Payback, which has been ongoing since 2010. This operation involves repeated DDos attacks against the web presences of prominent entrepreneurs and anti-piracy organizations that pur-sue copyright infringement. From a semiotic perspective, the conflict between the notion of the creative author, which is an effect of the Guttenberg Galaxy and generates property, and the understanding of free access to content in Cyberspace in the sense of the open-source movement manifests itself here.

With the expansion of authorship, the individual erodes as an epistemological monad and at the same time becomes the bearer of alternative understandings of the subject and an agent of resistance in Cyberspace: "Being Anonymous in the context of Wikileaks has a double promise: it promises to liberate the subject from the existing power structures, and at the same time it allows the exposure of these structures by opening up a space to confront them" (Bodó 2011, para. 2). In its organizational structure, Anonymous reflects the epistemology of the ephemeral, which Deleuze pointed out from an epistemological perspective with 'becoming liquid' and which Turkle updated in the context of the semantics of freedom in Cyberspace. It is important to note that Anonymous does not represent an acting subject but can instead be described as an effect of decentralized communication. Thus, "*Anonymous* [...] is by definition neither centralized nor limited to a certain number" (Bardeau and Danet 2012, p. 113, e.i.o.). Instead, Anonymous is characterized by "very flexible organizational structures" (Bardeau and Danet 2012, p. 114). Behind these organizational structures, the individual ('assignable self') disappears. This can be exemplified by the question of the identity of the hacker Topiary. Topiary was a spokesperson for the hacker collective LulzSec, which is associated with Anonymous. A 19-year-old suspect was arrested by British police forces in the Scottish Shetland Islands and transferred to London. However, the Web Ninjas group involved in the unmasking of LulZec expressed doubts as to whether the captured suspect was Topiary. The 'real' Topiary, on the other hand, was supposed to be a 23-year-old Swede:

> According to a *chat* between Topiary and another hacker, the arrested man is said to be an internet user who had misappropriated Topiary's identity in the past. The real Topiary then reportedly managed to get his counterpart arrested and decided to stay quiet until the storm passed. In any case, this story shows the difficulty for authorities in understanding the organizational mechanisms at work in *Anonymous* and stopping their actions. Whether Topiary is in prison or not, *Anonymous* has not stopped operating (Bardeau and Danet 2012, p. 116, ed.).

The decentralized organizational potentials of Cyberspace give rise to the possibility of civic identity being replaced by polyphonic communication in which the individual subject disappears: "The fact that one does not know the identity of individuals abolishes all prejudices regarding physical appearance or social or ethnic origin" (Bardeau and Danet 2012, p. 95 f.). To describe this dissolution of the individual subject through anonymity, it is possible to draw on Deleuze and Guattari's model of the organless body. In Deleuze and Guattari's epistemological diction, organs can be understood as subjectivizations through which individuals or bourgeois identities are produced. The organless body, on the other hand, represents a subjectless spatiality:

> An organless body is such that it can only be occupied and populated by intensities. Only intensities pass and circulate. Nevertheless, the organless body is not a scene, not a place [...] The organless body lets intensities pass. It produces them and distributes them in a *spatium* that is itself intense and has no extension. It is neither a space nor in space. It is matter that will occupy space to a certain degree – the respective degree corresponding to the intensities produced (Deleuze and Guattari 1992, p. 210, ed.).

The subjectless space of the organless body can be interpreted as a metaphor for Cyberspace: In the virtual world of Cyberspace, the individual subject gives way to the space generated by intensities or information exchange. These intensities, the exchanged information, constitute Cyberspace as such, which is not preceded by 'matter.'

Anonymous can be used to show how resistance becomes subjectless or a body without organs through Cyberspace: The virtual communication space of Cyberspace can be interpreted as a resistance space in which individuals become fluid or part of a digital swarm: Solidarization leads from the individual voice to a polyphonic chorus of resistance of the digital swarm. This digital swarm is organized without a center, quasi rhizome-like: "The digital swarm, unlike the mass, is not coherent in itself. It does not express itself as one voice" (Han 2013, p. 20). The spontaneity of cultural construction, which Han captures with the term hyperculture, and which is only made possible by the temporal-spatial implosion of the digital age, has significant analogies to the spontaneity of the digital swarm: The digital swarm is defined by "*fleeting patterns*" that have no "*fixed formation*" (Han 2013, p. 22, e.i.o.); it takes the form of an organless body. However, Han sees the lack of a 'fixed formation,' which the swarm does not exhibit, but the mass does, as a lack of agency. The mass as a supra-individual subject forms a 'we,'. The mass is now capable of 'common action.' The 'we' of the mass 'marches' (see Han 2013, p. 22), whereas the swarm is incapable of questioning power relations. According Han, the swarm, signifies the isolation of 'neoliberal economic subjects' (Han 2013, p. 24) who "do not form a we capable of acting together" (ibid.). In his description of the swarm, Han assumes a singular understanding of the subject. Accordingly, the mass assembles itself from a collection of individuals into an acting body. Since the decentralized swarm precisely does not form a body or a supra-individual 'we,' the swarm remains incapable of action as an organless body.

In contrast, Galloway and Thacker (2014), with reference to Deleuze and Guattari, point to the stability of networks that constitute the swarm's form of organization. Networks are characterized by the fact that they are reconfigurable: "[V]irtually this is what it means to be a network: to be capable of transformation, reconfiguration" (Galloway and Thacker 2014, p. 306). From the perspective of post-structuralist epistemology, the digital swarm can be understood as a dissolution of the individual subject. It is precisely from this dissolution of the subject, which can no longer be fixed and fitted into the symbolic order of traditional hierarchies and dependencies, that the digital swarm derives its subversive power as an organless body:

> The net is so powerful because it can react immediately and connect an unlimited number of people. As soon as someone feeds an idea, a piece of information into the net, it is instantly received. Other internauts can join in straight away, support the whistleblower who is defending himself against an injustice. The net is made for solidarity. We have passed on our wealth of experience to the young Egyptian revolutionaries and explained to them, for example, how they can protect themselves against tear gas (Ben Mhenni 2011, p. 44).

The Tunisian blogger Ben Mhenni (2011) sees the Internet as a subversive medium of communication against the background of her experiences during the 'Jasmine Revolt,' which led to the flight of the dictator Ben Ali in January 2011. The subject-less spontaneity of the digital swarm criticized by Han is seen by Ben Mhenni as a possibility of resistance. Fixed cultural spaces get caught up in a discourse made possible by the digital swarm. An aspect that Mishra (2017) also points out:

> Earlier, women were denied entry to certain places of worship, both in Islam and Hinduism. Although the battle for gender equality has been fought at various levels, including the court of law, it is finally the social media that has succeeded in forming a wider public opinion nationwide, which ultimately has resulted in gender equality in places of worship: It is media in general and digital media in particular that launched a continuous campaign for women's temple entry and succeeded in engineering a favourable public opinion for putting an end to such inequality in matters of worship (Mishra 2017, p. 129).

Ben Mhenni's and Mishra's arguments for a subversive effect of the Internet or the digital swarm stand in sharp contrast to Han's reflections on the neoliberally isolated individual. According to Han, the latter reproduces his powerlessness performatively by participating in the digital swarm since the swarm is incapable of action. These different interpretations of the digital swarm show the openness of the Internet, in which different digital cultures unfold.

Against the background of Deleuze's epistemology, the question arises as to whether it is precisely the transcendence of the subject in the organless body of the digital swarm that enables forms of subversion of power relations in the material-physical world. The resistance formulated in Cyberspace would then have an effect on the symbolic order of the material-physical world. To discuss this thesis, the subversive effect of digital communication will be elaborated in the juxtaposition of Spivak's model of the subaltern with the polyphonic and poly-directional possibilities of Web 2.0.

Spivak captures the – cultural – subjugation of the individual with the term subaltern. Subalterns are defined by the absence of a voice. Others speak for subalterns so that subalterns have no voice of their own in the interpretative patterns of hegemonic narratives. The digital swarm's decentralized structure makes it possible to break up hegemonically structured cultural spaces. Marginalized actors can step out of their state of subalternity by giving themselves a voice via digital media. This voice does not go unheard in a spatially limited cultural space but finds its resonance in the unbounded world of Cyberspace. The following subchapter will develop this thesis of overcoming subalternity through digital media.

## 2.3.3  From Subalternity to the Voice of the Digital

Spivak's 1988 essay "Can the Subaltern Speak?" is considered a "founding document of postcolonial studies" (Castro Varela and Dhawan 2015, p. 152) and "is among the most cited essays in contemporary humanities" (Castro Varela and Dhawan 2015, p. 193). Its central point of departure is the hegemonic structure of cultural spaces that prefigures cultural actors' narrative topoi. In this way, forms of voice and thus protest is rendered impossible. The starting point of Spivak's analysis is a critical examination of a conversation between Deleuze and Foucault: The two post-structuralist thinkers formulate the thesis that intellectuals do not need to act as advocates or 'voice leaders' for marginalized groups or the 'masses.' Such a proxy function would distort the positions of the social groups for whom they are supposed to speak. Instead of taking up the voice for marginalized groups, it would be necessary to listen to them. Deleuze and Foucault (1977) reject 'speaking for others' through the voice of intellectuals. Spivak criticizes Foucault and Deleuze for implicitly assuming social actors or individuals who can speak for themselves sovereignly. According to Foucault and Deleuze's logic, these social actors are aware of their subjection/oppression and, at the same time, possess the means to criticize the state of their subjection. Spivak develops the counter-thesis that such an understanding of social actors and their grasp of voice is not possible because there is no discursive space for these actors. Rather than asserting such a discursive space, the task is to trace the mechanisms by which individuals are fitted into hierarchies and processes of dependency and thus 'fall' through the discourses. Spivak exemplifies this speechlessness of actors by using the Indian woman as a gendered subject produced as a discursive effect of cultural processes of interpretation. Simplified and schematized, Spivak's reflections can be understood as the continuation of a postcolonial strategy of analysis, whose significant theoretical considerations are associated with the names of Said and Bhabha in addition to Spivak: With the approach of Orientalism, Said (see Said 2009) elaborates the discursivation of colonial, imperial politics and culture. In this process, the culturally foreign – such as the image of the Orient – is first discursively produced from a Western perspective. Dialectizes: In the third space, cultural identities are produced through a 'play of differences' consisting of movements of demarcation and acts of identification. In the process, a liminal hybridization of culture occurs. Spivak points out that in this third space of cultural representational struggles, all actors have a voice. According to Spivak, subalterns are actors which are functionalized for these representational struggles without these actors themselves having a voice. To clarify this consideration, Spivak draws on the concept of the subaltern as used by the South Asian Subaltern Studies Group in the context of postcolonial theorization and originally formulated by Gramsci.[16]

Gramsci formulated the concept of the subaltern in his prison notebooks, which he wrote during his imprisonment from 1929 to 1935 under the regime of Italian fascism. The term subaltern comes from the Latin (subalternus) and can be translated as 'subordinate,' 'of lower rank.' It was used in the Italian military to refer to officers of low rank or' subordinate officers' (see Castro Varela and Dhawan 2015, p. 186). In the context of postcolonial discourses, subalterns are defined by the fact that they have no voice in cultural discourses or negotiation struggles. Subalterns disappear from the analytical gaze. In other words: subalterns are consequently not recognized as political actors or as acting actors. The discursive invisibility or inaudibility of subalterns requires that the identification of subalterns is to be understood as a constructive act: Subalterns are not 'there' but need to be reconstructed. Spivak illustrates this with two examples relating to the discursive figure of the gendered subaltern Indian woman: The suicide of a woman resistance fighter and the practice of widow burning:

---

[16]The South Asian Subaltern Studies Group is an association of historians that has been publishing "Writings on South Asian History and Society" since the 1980s. These publications are intended, among other things, to give the history of the subaltern a place in postcolonial discourses.

The Hindu widow mounts the funeral pyre of the dead husband and sacrifices herself on it. This is the widow sacrifice. (The common transcription of the Sanskrit word for widow would be *satì*. The early colonial British transcribed it as *sutee*.) The rite was not practiced universally, and it was not fixed by caste or class. The abolition of the rite by the British was widely understood as a case of 'white men saving brown women' (Spivak 2008, p. 80 f., e.i.o.).

In her analysis of widow burning, Spivak draws on Vedic sources and Hindu texts such as the Rigveda and the Dharmashastras.[17] The Dharmashastras also address culturally sanctioned suicide. Through suicide, the culturally coded self-determination of the subject can be metonymically established: Suicide can be understood as a radical form of self-disposal. Man is free when he, as a finite being, can determine his end. From this perspective, Camus defines suicide as a central philosophical issue: "There is only one severe philosophical problem: suicide. The decision whether life is worthwhile or not answers the fundamental question of philosophy" (Camus 1999, p. 10). The degree of self-determination or self-disposal granted can be seen in the form of the cultural thematization of suicide. The widow's sacrifice does allow the wife a specific form of suicide. However, through a widow's sacrifice in this act, the wife symbolically does not primarily end her own existence but sacrifices herself *for* the husband. Without the husband's previous death, suicide is not culturally coded – although wives are "permitted to sacrifice themselves on the funeral pyre of the dead husband" (Spivak 2008, p. 86). As a sacrifice, suicide serves to solidify patriarchal structures: rather than self-determination of the subject through suicide, suicide allows us to see how much her relation to the husband defines the wife. Only through the death of the husband can the wife kill or sacrifice herself. The widow's agency is enabled through her relation to the husband. The practice of widow sacrifice comes under a struggle of interpretation due to English colonial policy in India. In the course of this interpretive struggle, widow burning is recoded from a colonial perspective: "In the discursive abrogation [...] of what the British perceived as a pagan ritual into what the British perceived as a crime, one diagnosis of free will was replaced by another" (Spivak 2008, p. 88): The practice of widow burning becomes evidence that Indian society is culturally in a primitive and barbaric state. "The image

---

[17] The Rigveda as one of the four Vedas represents the oldest, several thousand years old text collection of India. The Vedas, which can be translated from Sanskrit as knowledge/sacred teachings, represent a collection of texts that were initially passed down orally and later written down, and are among the sacred scriptures of Hinduism. Dharmashastras are a collection of Hindu law texts written in Sanskrit and deal with religious-legal regulations and were codified between the seventh and second century BC.

of imperialism as the founder of the good society bears the mark of advocacy [...] for women as *objects of* protection from their kind" (Spivak 2008, p. 84). Against the backdrop of an evolutionary concept of culture, colonization appears as a legitimate approach to counter the barbarism of primitive culture. In this context, the interpretation of the free will of the woman who wanted to make the widow sacrifice became the central point of 'discursive contestation':

> Of course, widow self-sacrifice was not an *immutable* ritual requirement. However, once the widow had decided to go beyond the ritual's letter in this way, any turning back meant a transgression, for which a special kind of penance was prescribed. In the presence of a local British police officer who supervised the sacrifice, on the other hand, it was considered a sign of genuinely free choice, a choice of freedom, if a woman, having made the decision, turned back from it [...] In the case of widow self-sacrifice, the ritual is redefined not as a superstition but as a *crime* (Spivak 2008, p. 88, ed.).

The colonial crackdown on widow burning, which was banned in 1829, becomes a metonymy for a claim to cultural superiority on the part of the British colonial power. While colonial and colonized cultures are constructed in constitutive demarcation, the treatment of widows becomes a forum for cultural struggles. Significantly, the body of the gendered subject becomes enclosed in the cultural discourses. The discourses take possession of the body and speak for the individual, who thereby loses his voice: A self-determined articulation of one's needs is given no discursive space, as Spivak points out with another example that comes from her family sphere: "As a young woman of 16 or 17, Bhuvaneswari Bhaduri hanged herself in 1926 in her father's modest home in north Calcutta. The suicide puzzled, since, especially since Bhuvaneswari was menstruating at the time, it was not a case of illicit pregnancy" (Spivak 2008, p. 104). Bhuvaneswari Bhaduri waited until her menstruation to commit suicide so that she could not be accused of illicit pregnancy, which otherwise often gave rise to suicide in such cases. This fact also points out that it was relevant for Bhuvaneswari Bhaduri to have interpretive power over the meaning of her death. Since pregnancy could not be assumed, Bhuvaneswari Bhaduri's suicide puzzled and challenged further attempts at an interpretation: "One attempted explanation saw her incomprehensible act as rooted in a possible melancholy, brought on by her brother-in-law's repeated taunts that she was too old to be still married" (Spivak 2008, p. 104). The mystery surrounding Bhuvaneswari Bhaduri's suicide did not unravel until nearly 10 years later when it emerged that "she was a member of one of the many groups involved in the armed struggle for Indian independence" (ibid.). She received the order of a political assassination, which she was unable to carry out: "Bhaduri's suicide was thus a complex orchestration that was not understood, in this sense Bhaduri could not speak" (Kerner 2012, p. 105).

In summary, subalternity, or the condition of subalternity, is defined by the fact that actors do not have a voice or discursive space and cannot be identified as 'authentic subjects' in the political struggle. Conversely, by speaking in a way that is heard – unlike in Bhuvaneswari Bhaduri's suicide – subalterns acquire subjectivity: consequently, they are no longer without a voice, no longer voiceless, and thus can no longer be identified as subalterns. From an emancipation theory perspective, the challenge arises to enable a discourse space in which subalterns can give themselves a voice. Spivak's examples, which serve as explanations for her thesis of the subaltern's speechlessness, highlight the extent to which individuals are enclosed in cultural discourses. It is this cultural enclosure of individuals that prevents the constitution of an emancipatory discursive space. Spivak's examples highlight the extent to which the impossibility of speech is tied to the cultural enclosed individual: widow burning focuses on the individual woman whose social function is defined by the culturally coded scope of action. If it has been exemplarily demonstrated here how speaking is prevented by the enclosure of the individual, this enclosure of the individual can be transcended through the polyphonic possibilities. Through their poly-directional and polyphonic orientation, Web 2.0 media can weaken the cultural enclosure of the individual by opening up supra-individual spaces of discourse for the individual.

Also, with reference to India, Mishra (2017) elaborates the articulation potentials of digital media for subalterns. The starting point of Mishra's analysis is India's culturally coded 'caste-based-inequality': "Inequality has manifested itself in Indian society in myriad ways at different points of time" (Mishra 2017, p. 125). Media change challenged these culturally coded hierarchies and relations of dependency:

> The explosion in digital communications – mobile phones, internet access, and digital cameras – is allowing citizens to engage in a public debate on a level unparalleled in history [...] Over the ensuing decade or so, it appears that groups once heavily underrepresented in their use of digital media – women, people of color, those in rural areas, and the poor – have greatly expanded their use of these new technologies (Mishra 2017, p. 127).

Digital technologies do not merely provide individuals with an instrument of articulation. Mishra attests to digital media the potential "to influence people especially the way they think about themselves and the world around them" (Mishra 2017, p. 127). Through the decentralized communication possibilities, the normative structure of the caste system can be questioned. Mishra analyzes how power structures come under critique through a digital-based media public sphere. The low-threshold usability and the participatory dimension of digital media enable a medial basis for the unfolding of the subversive impact of petite récits in the sense of postmodern epistemology and subversive diversity. In his media-theoretically oriented analysis, Mishra identifies four factors that constitute the participatory dimension of digital media or social media:

- One characteristic is the "inherent inclusiveness" (Mishra 2017, p. 129): "That is, this domain is devoid of all forms of exclusions and segregations. It is no longer restricted to aristocrats" (Mishra 2017, p. 126). Digital media are accessible and usable at a low threshold so that they can have an inclusive effect. Actors who are otherwise excluded from discourses are given a technical infrastructure and a discourse space through digital media.
- Mishra identifies "the property of openness" as the second factor (Mishra 2017, p. 126). Digital media are open to content of any kind: "Unlike the conventional media whose content development which is restricted to only few experts, the internet world does not impose any editorial scissors on the contents of the users" (Mishra 2017, p. 126).
- The third factor of the Internet's inclusive opening consists of the potentially hierarchy-free structure of the Internet. Mishra actualizes topoi the freedom semantics that have increasingly accompanied the discourse on Cyberspace since the 1990s: Beyond traditional hierarchies, dependency relations, and hegemonic structures, Mishra understands the Internet as a 'barrier-free zone': "Unlike caste system, digital media is a barrier-free zone devoid of disability. It is free from all forms of hierarchy, division, restrictions, untouchability, segregation and segmentation" (Mishra 2017, p. 126).
- The fourth factor is the ubiquity of the Internet: "Four, the property of universality. Today no part of the globe remains untouched digitally. In some way or the other, it has begun to engulf the whole universe and impinge upon people's everyday existence" (Mishra 2017, p. 126). The spatial opening through digital media also opens up cultural spaces: the formulations do not fade into a geographically limited space. Still, they can find ageographic resonances in the organless body of Cyberspace.

These four factors point to the polyphonic opening and participatory potential of digital media, which Reichert (2013) also points out concerning the Web 2.0 medium YouTube:

The transformation from a passive media consumer to an active actor who appropriates the screen as an interactive medium is based on two essential components: *Accessibility* and *Usability*. Accessibility can be understood as making information or technology accessible to all possible users, regardless of technical requirements and limitations. *Usability* refers to the effort to structure information clearly and concisely to enable the efficient use of data and technologies (Reichert 2013, p. 146 f., e.i.o.).

If a significant feature that characterizes the inclusive dimension of digital media lies in its usability, the emergence of Web 2.0 has led to an inclusive opening of the Internet. Besides the fact that everyone at least potentially has a digital voice, digital media's subversive power consists of a digital counter-public. This counter-public provides a voice to potential subalterns, defined by the fact that they have no discourse spaces. The enclosure of individuals in social hierarchies and relations of dependency makes it possible to fit them into cultural contexts, to functionalize them and thus render them voiceless. Digital media open up these repressive closed cultural spaces through their counter-speech: "Social media obtain, expose and publicize issues of social inequality in a barrier-free manner and provide a platform for participation in public issues" (Mishra 2017, p. 128). The poly-directional nature of digital media means that social software such as Twitter and Facebook enable a space for action for marginalized groups and individuals. "With the advancement in ICTs, like Facebook and Twitter, it has become very easy to capture events in android mobiles and send them to broadcast channels for greater salience" (Mishra 2017, p. 128). The digitally opened space of discourse constitutes a counter-public sphere.

The change in the media leads to a digital-based communication culture in which interpretive sovereignties can be doubted or questioned. Traditional cultural orders are undermined by the digital-based voice of minorities. The decentralized structure of the Internet and the transcendence of spatial restrictions lead to forms of resistance that overcome a repressive focus on the 'assignable self': "[D]igital media helped turn individualized, localized, and community-specific dissent into a structured movement with a collective consciousness about both shared plights and opportunities for action" (Howard and Hussain 2013, p. 25). A digital communication culture can subvert hegemonic interpretive claims that characterize metanarrations.

The decentralized, polyphonic, and poly-directional communication structure of Web 2.0 opens up a dialogical communication culture of petite récits realized through dialogical plurality and critical questioning. The efficacy of such digital-based communication unfolded in the so-called 'Facebook Revolution.' A central moment that led to the escalation was the attempt to deprive digitally formulated resistance of its discursive voice: In June 2010, 28-year-old blogger Kahlid Said was dragged out of an Internet café by Egyptian plainclothes police officers and beaten to death in the open street. Allegedly, Said had circulated a video on the Internet of police officers distributing confiscated drugs during a raid days earlier. Said's death was not investigated further, and it was claimed he choked on a packet of marijuana he had hidden under his tongue. Said was to be made a subaltern. The interpretation that was to be enforced by the state criminalized Said and aimed at morally delegitimizing him. The struggle for interpretive power manifests itself in a tragic symbolism in Said's death: the digital infrastructure gave resistance a voice. This was stifled by killing Said and at the same time attempting to discredit Said. However, this act of repression, in turn, became a digitally disseminated media event: photos of Said's abused body circulated on the Internet. Thousands of people protested in the streets of Cairo and Alexandria (Said's hometown). And as the demonstrations escalated, Said increasingly became a symbol of Egyptian police violence's arbitrariness, represented in the Facebook group 'We are all Kahlid'. As prophesied by The Mentor in the 1986 hacker manifesto, the individual can be accessed, but the organless body's plurality cannot be rendered voiceless. Through the digital swarm, Said possessed a symbolic voice even after his death. The use of digital media remained a central part of the resistance: protests were organized via Facebook, a counter-public was realized through digital media. At the same time, Egyptian state television showed images of an empty Tahir Square, Al-Jazeera broadcast video clips of protests in Tahir Square that the channel received by e-mail from Egyptian protesters. The user-centeredness of digital media had a subversive effect in resistance contexts. The mobile Internet flanked this subversive effect: "While before the upheaval only just under a quarter of the population had Internet access, more than two-thirds of all Egyptians owned a mobile phone. Thus, information about the protests was also distributed by phone or by collective SMS" (El Difraoui 2011, para. 3). The digital media of Web 2.0 and the mobile Internet are given the potential to break down hegemonic regimes of knowledge. As Howard and Hussain (2013) point out regarding the protests in the course of the Arab Spring: "[D]igital media provided the important new tools that allowed social movements to accomplish political goals that had previously been unachievable" (Howard and Hussain 2013, p. 18).

Cyberspace makes it possible to conceptualize the Internet as a space of freedom in which identity patterns become 'fluid,' individual resistance is transcended to a resistance of the digital swarm, and subalterns are given a voice. Assange (2012) updates this narrative when he contrasts a dialectical tension between Cyberspace and the material-physical world:

Once upon a time, in a place that was neither here nor there, we, the constructors and citizens of the young internet discussed the future of our new world. We saw that the relationships between all people would be mediated by our new world and that the nature of states, which are defined by how people exchange information, economic value, and force, would also change. We saw that the merger between existing state structures and the internet created an opening to change the nature of states (Assange in Assange et al. 2012, p. 2).

Following the tradition of anarchist theory, states are interpreted as structures of domination: "First, recall that states are systems through which coercive force flows" (ibid.). Assange acknowledges that states can have a democratic appearance when factions that are part of the state's power apparatus conflict with each other: 'Factions within a state may compete for support, leading to democratic surface phenomena' (Assange in Assange et al. 2012, p. 2). However, these democratic processes remain on the 'surface,' while the deep structures of state power fix fundamental hierarchies and dependencies: "Land ownership, property, rents, dividends, taxation, court fines, censorship, copyrights, and trademarks are all enforced by the threatened application of state violence" (Assange in Assange et al. 2012, p. 2 f.). From this perspective, the image of the free space of Cyberspace has discursive relevance in negotiating the possibilities of the Internet: Beyond the historical accuracy of this origin story, Cyberspace points to a potential space of freedom. Out of the freedom of Cyberspace, resistance is forming with Anonymous and other actors which has no 'name.'

The swan song to Cyberspace, which is being intoned about the Internet of Things, among other things, and which assumes that the virtual world and the material-physical world merge into one another,[18] appears premature from this perspective. In such a discourse, the libertarian implications of Cyberspace are excluded. Cyberspace becomes an imaginary space and a projection surface, a metaphor for the Internet's potential for freedom.

---

[18] "The distinctions of on- versus offline [...] or even 'real' world versus 'virtual' world of the Internet, which arguably made sense not so long ago, actually as well as the talk of their interpenetration, are so obsolete that even the reference to them has a historical character" (Jörissen 2017, para 1.).

## 2.3.4  'There Is a War on' – 'The Clash of Civilizations' on the Internet

The discursive dichotomization 'freedom space of cyberspace versus individuation compulsion of the material-physical world' leads to a thematization of the technical dependence of the 'platonic world' of Cyberspace on the material-physical world: "The platonic nature of the internet, ideas, and information flows is debased by its physical origins. Its foundations are fiber optic cable lines stretching across the ocean floors, satellites spinning above our heads, computer servers housed in cities from New York to Nairobi" (Assange in Assange et al. 2012, p. 2). In this juxtaposition, Cyberspace appears as the world of the mind confronted with the violence of the powerful of the material-physical world: "Like the soldier who slew Archimedes with a mere sword, so too could an armed militia take control of the peak development of Western civilization, our platonic realm" (Assange in Assange et al. 2012, p. 2).

According to this logic, the freedom space of the Internet, the world of the mind, faces a threat from the claims to power of the powerful in the material-physical world. These not only possess the power of disposal over the material-physical infrastructure of the Internet, but they also threaten to gain control over the 'platonic free space' of Cyberspace:

> The new world of the internet, abstracted from the old world of brute atoms, longed for independence. But states and their friends moved to control our new world — by controlling its physical underpinnings. The state, like an army around an oil well, or a customs agent extracting bribes at the border, would soon learn to leverage its control of physical space to gain control over our platonic realm (Assange in Assange et al., 2012, p. 2).

Assange warns of a transformation of the Internet from a space of freedom to a surveillance society: "If we do not, the universality of the internet will merge global humanity into one giant grid of mass surveillance and mass control" (Assange in Assange et al. 2012, p. 6). From this perspective, there is a 'freedom struggle' raging over the Internet, in which 'increased communication' is opposed to 'increased surveillance': "Increased communication means you have extra freedom relative to the people who are trying to control ideas and manufacture consent, and increased surveillance means just the opposite" (Assange in Assange et al. 2012, p. 21). The topoi of this argument stem from the traditional semantics of freedom of the libertarian cyberspace theorists. According to this narrative, a culture of emancipation and the disciplinary strategies of a society of control are dichotomously opposed. Accordingly, Holze (2017) holds that "[t]here is [...] war on the Internet" (Holze 2017, p. 96). This 'war' is fighting over interpretive sovereignties and forms of legitimation. The focus is on questions such as:

- What data, facts, information can be counted as knowledge?
- When is information' fake news'?
- Are there alternative facts? and
- Are whistleblowing and leaking, or sharing unauthorized data via digital platforms, strategies of legitimate knowledge communication, resistance, or betrayal?

In the course of this 'struggle,' the Internet threatens to be transformed into a surveillance technology, which Kammerl (2017) assesses as "threats to social sovereignty" (Kammerl 2017, p. 44). These threats have become "increasingly evident in recent years through reports of the Prism and Tempera wiretapping programs" (ibid.). Through these wiretapping programs, among others, "US and British intelligence agencies, respectively, were able to systematically siphon off digital traffic accumulated in Big Data" (ibid.). The Internet enables big data collection and measurement of the individual. Zimmermann (1990) makes this point using Google as an example, "Just look at Google. If you're a standard Google user, Google knows who you're communicating with, who you know, what you're researching, potentially your sexual orientation, and your religious and philosophical beliefs" (Zimmermann in Assange et al. 2012, p. 51). From this perspective, state surveillance and forms of data collection through digital mass media become blurred:

> As is now known, there are continuities in personnel and similarities in content between the two areas. In 2010, for example, Facebook's Chief Security Officer moved directly to the NSA. Such personnel castling take place at all levels and are facilitated not least by the fact that the prerequisites of the two areas are almost the same: exclusive access to vast amounts of data for the real-time analysis of social interactions (Stalder 2016, p. 234 f.).

It seems paradoxical that this monitoring of social interactions is realized through Web 2.0 media, which at the same time enable a postmodern form of emancipatory, dialogical communication. Web 2.0 media are mainly applied through so-called SNS. The starting point is the self-narration of individuals on the Internet, which produces data. Social visibility and thus the fixation of the individual is generated in this way. This gives rise to possibilities of panoptic surveillance. Thus, the web of SNS or the.

Social Web has become the most essential source of data for producing knowledge of government and control. Political control of social movements has also shifted to the web, as sociologists and computer scientists work together to produce a *riot forecast*, accessing the collected textual data from Twitter streams (Reichert 2014, p. 10, e.i.O.).

Howard and Hussain (2013), using the example of the Facebook Revolution in Egypt, point out that SNS or media such as Facebook and Twitter, which enable network-like swarm organization, simultaneously enable state access to protesters. SNS profiles can be used to identify protesters, as happened after the first protests organized via Facebook and Twitter: "A few days later, the Egyptian security services began using Facebook and Twitter as a source of information for a counterinsurgency strategy. They used social media alerts to anticipate the movements of individual activists (Howard and Hussain 2013, p. 22). By fixing the individual via SNS, surveillance access to the individual becomes possible. The former Wikileaks employee Applebaum (2012) sees an interlocking of state power and Internet companies in surveillance: "And this is the blurring of the state and corporation. This is probably the most important thing to consider here, that the NSA and Google have a partnership in cyber-security for US-national defense reasons" (Appelbaum in Assange et al. 2012, p. 55). This repressive perspective on the individual subject to surveillance by corporations and state power is flanked by a perspective of digitally controlled self-disclosure. A digital device such as the smartphone provides the equipment with which.

everyone can set up their mobile and portable one-person mini-panopticon in a self-constructed manner [...] The observation, matching, and further processing of individual synoptic initiatives' volatile distribution then again requires professional personnel. However, it is the 'users' of Google or Facebook. They fill the 'databases' in the course of their scattered, seemingly autonomous activities, but coordinated by the synopticon (Baumann in Lyon and Baumann 2014, p. 95).

The control dimension opens up a new analytical perspective on digital cultures: beyond the dialectic between Cyberspace and the material-physical world, an inscription of subject conceptions or neoliberal culture into the virtual world of the Internet can also be identified. These inscriptions manifest in a condensed form in the cultural practices of the SNS universe. Beyond all differences, the cultural practices of postmodern protest in Cyberspace and the neoliberal self-optimization of the entrepreneurial self (see Bröckling 2013) in the SSN universe are both characterized by an activist-volatile version of the individual: As part of a protesting network, a polyphonic, subversive diversity is constituted in the swarm. The collaborative dimension of dialogical communication processes produces (digital) content. *Digital Cultures rely on content generation and are based on the concept of the producing, acting individual.*

- Such an activist concept of the individual is embedded in the context of post-modern protest and articulated with narrative topoi of the cultural change of the protest movements of the 1960s and 1970s.
- Within the SNS universe, on the other hand, the activist individual is framed in neoliberal narratives, which also selectively refer to a postmodern understanding of the subject, but at the same time fits this understanding of the subject into the metanarrative of a neoliberal worldview.

This thesis will be developed in the following.

# The SNS Universe of the Control Society

3

## 3.1 From Postmodernism to Neoliberalism

### 3.1.1 The New Spirit of Capitalism

Postmodernism's cultural discourse constituted a space for an aesthetic of self-expression or an aestheticized individualism, defined by diversity and flexibility. These terms with positive connotations inscribed themselves in the discourses of (self-)understanding of capitalist thought and action. Against the backdrop of their analysis of management literature, Boltanski and Chiapello (1999/2013) conclude that the 'spirit of capitalism' changed fundamentally: "The management literature of the 1990s contains ideals, suggestions for employee organization, organizational modalities of the object world, and security guarantees so different from the 1960s that it is difficult to avoid the realization that capitalism has, to a large extent, changed its spirit in the course of the last 30 years" (Boltanski and Chiapello 2013, p. 142).

The protest movements of the 1960s and 1970s were also movements of cultural critique. The irony of the postmodern protest of the Spontis had anti-capitalist features and represented an anti-capitalist alternative culture. Against conventionalized patterns of authority, a left-wing alternative culture was set, which, through children's and book and neighborhood shops, residential communities, as well as through critical psychoanalysis (see Braun 1979), posited the image of a subject developing according to its own needs. Cultural critique and social critique were interwoven, lifestyle and critique of capitalism merged, and in their interconnectedness constituted a left-alternative and thus at the same time epistemological and power-critical small narrative in the sense of postmodern epistemology. In the course of the change in the spirit of capitalism, parts of this petite récits of

D. Kergel, *Digital Cultures*, https://doi.org/10.1007/978-3-658-35250-9_3

left-alternative critique of culture and capitalism became topoi of neoliberal (meta-) narrative. Neoliberal narratives they were, as it were, incorporated into the spirit of neoliberalism. Thus became

> Autonomy, spontaneity, mobility, disponibility, creativity, pluricompetence [...] the ability to form networks and to reach out to others, openness to the different and the new, the visionary gift, the feeling for differences, the consideration for one's history and the acceptance of diverse experiences, the inclination towards the informal and the striving for interpersonal contact - borrowed directly from the world of ideas of the 68ers (Boltanski and Chiapello 2013, p. 143f.).

When Boltanski and Chiapello mention, among other things, 'openness to the other and the new' as well as 'consideration for one's history and diverse experiences,' they describe an epistemological attitude that can be described with the term post-modern epistemology due to its epistemological implications vis-à-vis claims to totality. In the establishment of the new spirit of capitalism, this epistemological attitude is detached from social demands or critique of capitalism. This process of detachment opened up the left-alternative narrative topoi to a neoliberal world-view: "These themes, which in the texts of the '68 movement are associated with a radical critique of capitalism (especially exploitation) and with the proclamation of its imminent end, in a sense take on a life of their own in the literature of neo-modernism, forming goals in their own right" (Boltanski and Chiapello 2013, p. 144). From a critique of capitalism-orientated position, the detachment of indi-vidualist values of freedom from social issues leads to alienation. At the same time, the separation of individualist values of freedom from social problems leads to alienation, and it discursively enables a neoliberal framing of the individualist free-dom of the subject:

> The critique of the division of labor, hierarchy, and surveillance, that is, of the way industrial capitalism alienates freedom, is thus detached from the critique of alien-ation by commodity society, of oppression by impersonal market forces, with which it is nevertheless always associated in the protest writings of the 1970s (Boltanski and Chiapello 2013, p. 144).

From this perspective, neoliberalism can be understood as a concept of the discur-sive and cultural transformation that capitalism underwent due to the protest move-ments of the 1960s and 1970s. Through this transformation, capitalism received its 'new spirit' in which it pretended to be free and (neo) 'liberal': "In the 1980s, neoliberal ideas prevailed, partly because some of the values, procedures, and methods propagated by the New Social Movements were detached from their political context" (Stalder 2016, p. 33). The values of postmodernism were placed in neoliberal contexts, which also resulted in a restructuring of work culture:

> A gentleman of management consultants, restructuring experts, and new entrepreneurs began to promote flat hierarchies, personal responsibility, and innovation and set about transforming large companies into small, flexible units. Work and leisure were no longer to be separated, and all aspects of a person's life were to be integrated into work. The personal identification of each individual with their job was now considered a prerequisite for economic success in the new capitalism (Stalder 2016, p. 33).

In the course of this transformation process, the autonomy of the subject, became an integral part of the discursive self-thematization of capitalist thought: "As it seems to us, neo-management thus appears to respond to the two needs for authenticity and freedom that have historically been shared by the so-called 'artist critique', neglecting, in contrast, the problematic areas of egoism and inequalities traditionally associated in 'social critique'" (Boltanski and Chiapello 2013, p. 143).

The individual has held out the prospect of a "certain gain in freedom" (see Boltanski and Chiapello 2013, p. 134) through the new spirit of capitalism or neoliberalism. "In the new world, anything is possible because creativity, reactivity, and flexibility are considered the new buzzwords" (ibid.).

A so-called 'reordering of the social' actually euphemizes the dissolution of the social. This reordering of the social towards a neoliberal spirit leads to a culture of neoliberalism. Identity-patterns, understandings of self/world-relations are narrated in terms of neoliberal value-settings. The topoi of this narrative are taken from the narrative inventory of emancipative approaches, which can be located in the context of postmodern epistemology. The emancipative idea of empowering individual agency in a social context has been reinterpreted as a topos that addresses the individual's self-responsibility. The individual does not become ephemeral or 'fluid' in the sense of Deleuze. Still, it is characterized by flexibility and mobility.[1]

This results in the ideologeme "that everything that could be an obstacle to mobility must be eliminated" (Boltanski and Chiapello 2013, p. 171). Flexibility and mobility thereby correlate discursively with the reduction of job security through the elimination of permanent employment contracts, since indefiniteness would imply a form of boundedness, constraint, and fixity: "The imperative of unboundedness presupposes first and foremost a renunciation of stability and

---

[1] Flexibility and mobility as central topos of neoliberal narratives prove to be connectable to the mode of use and narration of the so-called Mobile Internet.

rootedness, of attachment to a place and the certainty of long-standing contacts" (Boltanski and Chiapello 2013, p. 169). Fixed-term employment contracts, a central feature of current precarization processes (see Kergel 2016), appear necessary to enable the individual's autonomy and flexibility. Fixed-term employment contracts, in turn, constitute the employment relationships that are absorbed into the work form of the project:

> Everyone is aware at the moment of participating in a project that the enterprise they are involved in will be of limited duration. One knows not only that it can come to an end at some point, but that it must come to an end at some point. The perspective of an inevitable and desired conclusion thus accompanies the *commitment*, but without diminishing the enthusiasm (Boltanski and Chiapello, p. 156, ed.).

As a temporary, targeted, and unique work, the project offers the framework in which the neoliberal individual can play out his or her flexibility, autonomy, and mobility and develop activism: "To be active means to bring *projects into* being or to join *projects* initiated by others" (Boltanski and Chiapello, p. 156, e.i.o.). In the 'magic word project' (see Bourdieu 2016, p. 7), the ambivalence of postmodern semantics of freedom is transferred into the neoliberal sphere:

> The *project to* which the subject liberates itself today proves itself to be a figure of coercion. It unfolds constraints in the form of performance, self-optimization, and self-exploitation. We are living today in a particular historical phase in which freedom itself gives rise to constraints. Freedom is *the* counter-figure of constraint. Now this counter-figure itself generates constraints. More freedom, therefore, means more constraint. That would be the end of freedom. So today, we find ourselves at an impasse. We can neither move forward nor backward (Han 2013, p. 65f., e.i.o.).

Project, autonomy, mobility, and flexibility stand for overcoming an authoritarian structure in the context of work relationships (see Boltanski and Chiapello 2013, p. 142). In this critique, "it is not difficult to discern an echo of the anti-authoritarian critique and the desires for autonomy that were expressed with vigour in the late 1960s and 1970s" (Boltanski and Chiapello 2013, p. 143). The freedom of postmodern epistemological critique is reinterpreted as a freedom of the neoliberal market: the individual is free to behave flexibly to the inescapable 'constraints' and dynamics of the neoliberal market.

## 3.1.2   Neoliberal Dominance Culture as Metanarrative

In the course of a neoliberal reorganization of welfare state systems, the self-understanding of a society or the symbolic order of society is challenged across fields, which effects the construction of narratives of precarization. Häcker (2011) points out that an "expansion of a neoliberal discourse of control [...] can be observed. This entails an absolutization of the logic of the market or a delimitation of the rationality of the economic system to other areas" (Häcker 2011, p. 172). Bourdieu problematizes this process as a 'political implementation' of the "utopia [...] of neoliberalism" (Bourdieu 1998, p. 109), which can be seen in an establishment of "the pure and perfect market, as demanded by the policy of deregulation of the financial markets" (Bourdieu 1998, p. 110). In this context, however, neoliberalism can be understood less as a self-contained model, since it has neither "erected a coherent edifice of ideas" (Bröckling 2013, p. 78), "nor does it have a unified political (or anti-political) practice" (ibid.): "The economic theories that operate under the self-chosen or externally ascribed label 'neoliberalization' are anything but homogeneous" (Bröckling 2013, p. 104). Rather, the "diverse currents" (Bröckling 2013, p. 78) of neoliberalism can be understood as a manifestation/ radicalization of competitive discourses that ascribe a supra-individual rationality to market events.

The secularization of the individual corresponds to an understanding of the economy that – committed to the bourgeois primacy of the rational – assumes a rational market. This can be exemplified by von Hayek's understanding of the market and society – a pioneer of neoliberal approaches. Hayek (1981) provides an interpretation of society which bases on terms of rational dynamics. Society "is not an acting person, but an ordered structure of actions that results from its members observing certain abstract rules" (Hayek 1981, p. 131). This understanding of rationally based automatism is applied to economic processes. The 'benefits of the market economy' cannot be intentionally produced by an individual. "This imposes on us the obligation to accept the results of the market even when it turns against us" (Hayek 1981, p. 131). Following Kant (see Kant 1784; Kergel 2011a, b), social processes are interpreted as processes of supra-individual rationality, whereby von Hayek constructs a short-circuit between society and economic action, which is also pointed out by Butterwege et al. (2008): "Thus, Adam Smith's one 'invisible hand' becomes, in Hayek, two invisible hands – the invisible hand of economic coordination and the invisible hand of social organization" (Butterwege et al. 2008, p. 58). Neoliberal thinking is characterized by a radical abandonment of regulatory intervention in the supposedly rationally unfolding processual dynamics of the market.

To ensure an appropriate unfolding of the market's rational logic, the principle of competition is needed. Here, neo-liberal thinking takes the state to the task to promote the principles of competition in a socially appropriate manner:

> If state legislation is to enforce the market laws, the public authority must not under any circumstances seek economic-social control over the economic process. Instead, it must exclusively provide a formal framework in which economic actors can pursue their individual goals as comprehensively as possible (Bröckling 2013, p. 83).

The state becomes an assistant to the market. The main task of the state is to guarantee the development of the market through competition. According to this logic, the individualism of bourgeois society experiences a revaluation since the individual functions as a monad of the market. The bourgeois concept of freedom is given a specific re-interpretation in the context of neoliberal programs. In this context, freedom always refers to economic freedom or freedom in a neoliberal market, which is free of state regulation. The meaning of the market takes on crucial importance in the establishment of neoliberal discourses and policies. The market is transformed from the forum in which individuals operate to an overarching principle of social practice. Paradigmatically, this can be seen in the public choice approach (see Buchanan and Tullock 1962): according to this, even in the sphere of politics, no collective interests are represented, but individual interests. Concerning the (welfare) state, this means that no general social interests are represented in this political sphere. Instead, the state serves particular interest groups. Accordingly, it is necessary to transfer neoliberal market logic to the welfare state field and, consequently, to other social areas – as a regulative to individual interests.

The anthropologizing model of homo economicus defines the human being as driven by rational choices to maximize individual utility. The model of the homo economicus can be understood as an effect of secular bourgeois society. Homo economicus establishes itself as a metaphor of a rational, activist orientation towards this world. Foucault (1974) points out that "since Kant, the question of finitude has become more fundamental" (Foucault 1974, p. 315). A potential nihilistic perspective, which are characterized by rational pragmatism, is countered by the surrogate of homo economicus:

> It is no longer the game of representation in which economics finds its principle, but it finds its origin in that dangerous area in which life confronts death. [...] The positivity of economics is located in this anthropological doctrine. *Homo economicus* is not the one who represents to himself his own needs and the objects that can alleviate

them. He is the one who spends, consumes, and loses his life trying to meet the threat of death (Foucault 1974, p. 315, e.i.o.).

This idea of man appears as a subtextually effective model of neoliberal theory formation and politics. The ideal image of a rationally reflecting and acting bourgeois individual is increasingly thematized and actualized in market logics. These neoliberal market logics formulate economic and consequently interpersonal relations as competitive relationships. The *homo economicus* becomes subtextually the anthropological blueprint of social interpretation of the meaning: Bourdieu thus points out that the "path to a neoliberal utopia [...] takes place within the framework of a transformational, or [...] *destructive* labor" (Bourdieu 1998, p. 110, e.i.o.). The form of this 'destructive labor' forms an "economic regime" (ibid.). It relies discursively on a "kind of logical machine that can be understood as a chain of constraints" (Bourdieu 1998, p. 111). This logic of argumentation unfolds a cross-field influence. The premise of neoliberal figures of thought consists of a cross-field automatism of market logic: Social dynamics regulate and take place according to the market's opaque but rational 'algorithms.' In the context of the market-centeredness of neoliberal approaches, a neoliberal metanarrative is discursively installed:

> In Great Britain and the USA, Conservative think tanks had already been calling for the market-radical turn since the 1970s. They postulated not the state's withdrawal, but the universal orientation of its interventions towards the establishment of an 'enterprise culture' – an activist program that was not to leave out any sphere of life (Bröckling 2013, p. 53).

The image of homo economicus is discursively ontologized and has been given an increasingly real-political meaning, especially since the 1970s. In 1973, for example, neoliberal approaches were implemented in Pinochet's military junta government in Chile. This was followed by the implementation of neoliberal policies via the "Reagan Revolution in the US and [...] [the] Thatcherism in the UK" (Biebricher 2012, p. 87) up to the 'Hartz IV reforms' in Germany. Since then, the implementation of these policies has increasingly shaped the self-understanding discourses of bourgeois society. In the course of establishing neoliberal positions, "all *collective structures are called into question*" (Bourdieu 1998, p. 111, h.i.o.), "which can put any obstacles in the way of the logic of the pure market" (ibid.). This process evokes a restructuring of the symbolic order of bourgeois society: "That everyone should become the entrepreneur of his or her own life was in the logic of Thatcherism and Reaganomics, which put individual self-responsibility at the top of the political agenda and flanked the dismantling of welfare state security systems with this postulate" (Bröckling 2013, p. 53). The result is a neoliberal re-structuring of Western societies. This restructuring evokes the metanarrative of neoliberalism. This metanarrative premise is a competitive orientation and a culture of economic dominance that extends to all social fields. Economic solidarity as a characteristic of freedom, which plays a central role in the course of left-alternative social concepts, is faded out in the context of neoliberal discourses. This omission of social solidarity makes the welfare state's neoliberal rollback appear a 'liberation process.' The neoliberal transformation significantly restructures the world of work and leads to an 'unrestricted rule of flexibility.' This flexibility and neoliberal freedom form the preconditions for responding appropriately to the dynamics of the market:

> And so the unrestrained reign of flexibility soars, one of fixed-term employment, temporary work, 'social plans', carries competition into the companies themselves, between independent branches, workgroups, ultimately of each against the other, which the *individualization of* employment relations brings with it: individual targets, individual evaluation procedures, individual wage increases or performance bonuses, individual promotions; strategies of 'delegation of responsibility' designed to ensure the self-exploitation of employees, employees who, while being in a strictly hierarchical relationship of dependence like simple wage-earners, are at the same time made responsible for their sales figures, their branch office, their business, like 'self-employed persons'; a 'self-control' that allows its 'inclusion' of employees to reach beyond the salaried workforce according to all the rules of 'participative management' – all rational techniques of subjugation that, under massive investment in labour, not only in the managerial sphere, ultimately vie to weaken or eliminate collective cohesion and collective solidarity (Bourdieu 1998, p. 112, e.i.O.).

One effect of these social transformations is the restructuring of the symbolic order of bourgeois society, which can be seen, among other things, in the reorganization/dismantling of the welfare state and also manifests itself in the fact that "social relations are increasingly permeated and shaped by economic calculations" (Häcker 2011, p. 172).

### 3.1.3   Entrepreneurial Self as Topos of Neoliberal Metanarrative

These cultural shifts of neoliberalism are analytically reappraised by Bröckling (2013). Bröckling analysis the neoliberal normative demands on the individual as manifestations of the "entrepreneurial self":

In the figure of the entrepreneurial self, both a normative image of a man and a multitude of current self-and social technologies are condensed, whose common vanishing point is the orientation of the entire way of life to the behavioral image of entrepreneurship [...] One is not an entrepreneurial self, one is supposed to become one. And one can only become one because one is always already addressed (Bröckling 2013, p. 47).

On the methodological basis of discourse-analytically oriented analyses, Bröckling identifies "some common basic chords" from the polyphony of neoliberal programmatic (Bröckling 2013, p. 106). For Reckwitz, the "period around 1970 and 1980" (Reckwitz 2006, p. 101) represents a [...] "structural as well as an epochal cultural threshold, which at the same time includes a transformation of the subject order" (ibid.). Regarding Bröckling's analyses, this new subject order can be understood as establishing the neoliberal subject. This establishment of the neoliberal subject order is linked back to neoliberal restructuring of the world of work:

In the field of work, this is a shift from the 'organization man' of white-collar culture to a combination of creative subject and 'enterprising self'; in personal relationships, there is a break from the now seemingly conventionalist 'peer society' to a subject that strives for 'self-growth' in its intimate relationships: analogously, in the field of consumption, a reorientation from socially copied consumption to individual aesthetic consumption becomes clear (Reckwitz 2006, p. 101).

The neoliberal individual is an individual "who chooses" (Bröckling 2013, p. 106). "The neoliberal concept of freedom avoids the drama of the existentialist self-design like the pathos of emancipation from social constraints and instead foregrounds the pragmatic choice between available alternatives" (Bröckling 2013, p. 106). This freedom of the individual is in turn embedded in the constellations of neoliberal self-optimization and insecurity. In terms of the rate of profit that defines economic growth, on the individual-theoretical level of neoliberalism, the "growth of the 'Me Incorporated' & Co [...] has no goal. Those who follow this goal will never arrive, but always remain in motion" (Bröckling 2013, p. 145). In doing so, "individuals [...] should maximize their power over themselves, their self-esteem and self-confidence, and their health, as well as their work performance and wealth" (Bröckling 2013, p. 61).

Performance and competition principles form the basis on which the individual neoliberal acts. At the same time performance and competition principles represent the cornerstones of an emerging neoliberal metanarrative: As an inescapable or unquestionable premise, the market is not defined by the freedom of critical questioning, as is postmodern epistemology, but by neoliberal freedom of action. A logic of freedom is constituted that establishes a 'sporting against each other', instead of a dialogical questioning of truth claims. According to neoliberal logic, the individual's freedom unfolds in a free market in which individuals compete against each other. From this perspective, the neoliberal concept of freedom is based on competition, which structures the free market. This competition presupposes the freedom and activism of the individual. The freedom of competition must be protected from intervening regulations – for example, on the part of the state: "Competition can only unfold its stimulating effect if it is not overridden by interventions that prevent or distort competition" (Bröckling 2013, p. 106f.). If the unfolding of the free market is given, the individual can unfold his potentials through neoliberal competition. In this, "[t]he neoliberal apologies of competition are thoroughly Darwinian in the foundation" (Bröckling 2013, p. 97). Competition dissolves the social as solidary practice. Thus, it is "a dynamic event. In this perspective, the market does not appear as a place of peaceful reconciliation of interests using exchange, but as a confusing sequence of gaps that open up and close again" (Bröckling 2013, p. 107). According to Bröckling, the entrepreneur is characterized by "recognizing and exploiting these gaps" (ibid.). Neoliberal freedom is uncritical freedom that keeps the individual in a permanent state of tension through the logic of competition.

### 3.1.4   Experiences of the Precariousness of the Entrepreneurial Self

The neoliberal logic of competition defines the individual as fundamentally threatened by the claim to freedom of others. The postmodern moment of reciprocity is thus abolished. The other becomes a symbol of constant threat: "Because everyone can always assert his position only for the moment and to his competitors, no one may rest on what he has once achieved" (Bröckling 2013, p. 72). This implicitly suspends ethics of compassion in the sense of Lévinas. Neoliberalism knows no ethics: from this perspective, the entrepreneurial self is defined by a boundary between it and other entrepreneurial selves. Others become a threat. This threat spurs constant activism so that individuals become 'entrepreneurs in their own right':

> The constant fear of not having done enough or not having done the right thing and the unstoppable feeling of inadequacy are as much a part of being an entrepreneur in one's own right as mercantile skill and the courage to take risks. Self-employed people are called that because, firstly, they work themselves and, secondly, they work all the time, says a famous bon mot among "Ich-Ag's" [Me Incorporated]. No amount of effort can guarantee security, but failure is a certainty for anyone who lacks toughness towards themselves (Bröckling 2013, p. 74).

The flip side of neoliberal freedom and the entrepreneurial self is the fear of the precarious individual: "The precariatized mind is fed by fear and is motivated by fear" (Standing 2011, p. 20). Marchart points out that precarization has 'objective' and 'subjective' dimensions (Marchart 2013, p. 13): 'Objectively, precarization turns out to be a form of post-Fordist regulation that undermines the Keynesian welfare regime' (Marchart 2013, p. 13). As a consequence, "employment security and regularity of income [...] are called into question across the board" (Marchart 2013, p. 13). According to Marchart, the experience of these transformation processes leads "on a subjective level to the anxiety neuroticism of the individual" (Marchart 2013, p. 13). The experience of neoliberal freedom as 'stable instability' (Kergel 2016), which appears to be necessitated by the movements of the market, evokes a 'precarized habitus' (Bourdieu 1998, p. 112) or a habitus of the precarized individual. This habitus is defined by a tension that spans between the poles of self-empowerment and experiences of anxiety. This tension, which is constituted by neoliberal freedom, is exemplified in the competition between individuals, which the entrepreneurial self produces in its actions in the first place and at the same time makes possible. In this context, the entrepreneurial self is precarious since it is exposed to the stable instability that characterizes precarization and experiences this precarization fearfully.

### 3.1.5 'Hey You There!': Neoliberal Subjectification

Bröckling understands the establishment of the entrepreneurial self and thus also of the precarious habitus as total: "An outside untouched by the entrepreneurial subjectivation regime or an interior space of the self-withdrawn from it does not exist – or if it does, then only as a zone of future conquests where unused resources await their development" (Bröckling 2013, p. 285). This explanation of totality strengthens the interpretation of neoliberal discourse as a form of contemporary metanarrative. According to Bröckling, the metanarrative of the neoliberal, which is condensed in the metaphor of the entrepreneurial self, integrates forms of postmodern understandings of the subject. But the subversive 'becoming liquid' appears powerless. Thus, both "the liquefaction of positions" (Bröckling 2013, p. 285) and the "jumping back and forth between plural identities [...] does not lead out of the spell" (ibid.) of neoliberal interpellations and fixations:

> The nomadic, 'queer' or hybrid subjects that populate poststructuralist theories – from Gilles Deleuze to Judith Butler to Homi Bhabha – as empathically charged counter-calls may subvert the pressure to homogenize that is still effective in a

post-disciplinary society with a conundrum of blurred or shifting identity constructions. Still, they have little to offer in opposition to the flexibilization imperative of a
radicalized market economy (Bröckling 2013, p. 285).

The process in which neoliberal narratives are inscribed in the individual can be
analytically grasped with the concept of subjectivation. A constitutive element of
subjectification processes can be seen in interpellation analyzed by Althusser
(1977). Althusser exemplifies this process using an example that Bröckling understands as a "primal scene" (Bröckling 2013, p. 27):

> One can imagine this invocation following the pattern of the simple and everyday call
> by a policeman: "Hey, you there!" Once we assume that the imagined theoretical
> scene takes place on the street, the called individual turns around. By this simple
> physical turn of 180 degrees, he becomes the subject. Why? Because it acknowledges
> that the call was "precisely" for him and that it was "precisely he who was called"
> (and no one else). As experience shows, the practical telecommunications of invoca-
> tion practically never miss their man: whether by verbal acclamation or by a whistle,
> the called always recognizes precisely that it was he who was called. In any case, this
> is a strange phenomenon that cannot be explained by a "feeling of guilt" alone, de-
> spite the multitude of people who "have something to blame themselves for"
> (Althusser 1977, p. 142f.).

The interpellation can be described as a "holistic ideologization" of the individual
insofar as the invocation has a cognitive and emotional dimension. Following
Foucault, Butler formulates a concept of subjectivation that further differentiates
this inscription process of symbolic order into the individual through the analysis
of reproduction processes: "'Subjection' signifies the process of becoming subor-
dinated by power as well as the process of becoming a subject. Whether by interpellation, in Althusser's sense, or by discursive productivity, in Foucault's, the subject is initiated through a primary submission to power" (Butler 1997, p. 2).

Hierarchies often do not have a direct effect, but manifest themselves performatively in (linguistic) action, unnoticed by the actors: Interpellations are embedded
in social systems of meaning and conceal and/or legitimize the processes of hierarchization of the symbolic order. A central aspect of Butler's reflections on the constitution of power builds on the principle of repetition organized primarily in language. Using "stylized repetition" (Butler 1991, p. 206), interpellations sustainably
unfold their effects. It is precisely the infinite reproduction of interpellations
through repetition that gives "the performative utterance its binding or conferring
power" (Butler 1995, p. 297). Moebius (2016) elaborates that in "Butler's model of
performative power [...] power depends on repetition" (Moebius 2016, p. 169).

"[E]ven through the repetitiveness of discursive-normative instructional structures, power unfolds" (ibid.).

Interpellations obtain their power through their (linguistic) repetition, which can be grasped with the concept of the performative. Conceptually, Butler's approach can be understood as a performative-interpellative analysis of power. Since this form of power analysis foregrounds the dynamics of socialization processes, it can be read as an analysis of the productive, generating dimension of power. One effect of this productive dimension of power can be seen in the fact that the individual as a subject acquires agency: "The subject now presents itself as the double structure of a *subiectum*: by submitting to certain cultural orders that physically and psychologically 'inscribe' in him the characteristics of acceptable subjecthood, the individual can only form those competencies of self-government, expressivity, rational choice, etc. that are supposed to constitute a subject" (Reckwitz 2006, p. 78).

The subjectification model does not focus on emancipative self-reflection. Instead, the social inscriptions in the individual's understanding of the self/world are taken into epistemological view. The process of subjectification is conceptually analyzed through the model of interpellation, while Butler emphasizes the performative dynamics of the permanence of interpellation. Culler (2002) understands this process as one in the course of which the subject is conceived as decentered. This means a break with an understanding of the subject in which the subject is conceptualized as the autoreferential starting point of reality construction:

> Suppose it is the case that the possibilities of thought and action are determined by a set of systems that the subject neither controls nor comprehends. In that case, the subject is 'decentered' in the sense that it cannot be thought of as an origin or center to which one refers to explain actions. Instead, it is something that emerges under the influence of such forces (Culler 2002, p. 157).

The epistemological model of subjectivation can be opened up for sociological analyses. "With subjectification [...] a multiform process comes into view" (Rieger-Ladich 2012, p. 66). In this process, "not only norms and conventions, organizations and institutions are entangled [...], but also individuals, social groups as well as functionaries representing the state and its particular power to give a name" (ibid.). From this perspective, the subject is the effect of subjectification practices. The subject thus denotes "the social-cultural form in which the individual expresses himself, in which practice imprints itself in him; as a subject, the individual imbues himself with social-cultural criteria of subjecthood, he submits to them to be able to become active" (Reckwitz 2006, p. 95).

## 3.2    The Start-Up Project: Sharing Economy as an Alternative Business Culture

Through subjectification, the neoliberal metanarrative increasingly inscribes itself into social structures and shapes individuals' self/world relations. The Internet proves to be a predestined projection surface for neoliberal perspectivizations: The implosion of time and space can be read as a surface in which neoliberal flexibility can unfold. The independence of Cyberspace has the potential to realize a free market beyond nation-state borders. Through Web 2.0 media, content can be produced at a low threshold. This content can be offered for sale in the sense of neoliberal logic of exploitation. These possibilities of neoliberal action and the unfolding of neoliberal narratives manifest themselves paradigmatically in the discursive thematization of start-ups:

> Tech companies like Google, Apple, Amazon, and Facebook dominate the digital tech industry as it takes off, their brand names synonymous with commercial success. Start-ups with such bold names as Uber, Airbnb, Twitter, Dropbox, Upwork, TaskRabbit, Instacart, and dozens more are seen as beacons of entrepreneurial innovation and daring. They are hyped as the incarnation of the future (Hill 2017, p. 9).

Start-ups embody the creativity and artistic aesthetics of the new spirit of capitalism. This new spirit of capitalism is constituted in the wake of neoliberalism and finds in Cyberspace a medial sphere for the unfolding of neoliberal flexibility through the implosion of time and space. Expressions of innovation are the "idiosyncratic names such as Wooga, Clickworker, Appjobber, Foodzora, ZenMate, and Zalando, for example" (Hill 2017, p. 9).

Start-up's challenge established business models and their hierarchical and authoritarian structures. From the perspective of cultural theory, start-ups correspond to the approaches of the protest movements of the 1960s and 1970s as well as to the conceptions of the neoliberal flexibility of a free market: authoritarian structures and firmly established hierarchies are opposed to neoliberal flexibility. They are exposed to the free market by the attacks of the 'creativity' and 'innovative power' of capitalism's new spirit. In this context, "[t]he start-up economy [...] encompasses a large number of companies from diverse industries with diverse occupations" (Hill 2017, p. 36) and centrally relies on the model of the sharing economy. The creativity and innovation of start-ups are metonymically demonstrated in the actualization of the sharing economy. The sharing economy underwent a left-alternative contextualization in the aftermath of the protest movements of the 1960s and 1970s. Thus, "[t]he countless housing, work, cultural, and social projects

[...] saw themselves as counter-designs to the factory, nuclear family, and university and reacted not least to the failure of other political concepts ('march through the institutions,' 'armed struggle,' party-communist cadre organizations) after the upheaval of 1968" (Bröckling 2013, p. 257). In the course of this, the concept of "the unfolding of a system of counter-economy" (ibid.) also emerged. These alternative projects anticipated "the autonomization, responsibilization, and sustainability programs" (Bröckling 2013, p. 259) that characterize the spirit of the new capitalism and have "penetrated all pores of society since the 1990s at the latest" (ibid.).

With the sharing economy, a business model was simultaneously prefigured that plays a central role in start-ups' platform concept. The infrastructure required for the sharing economy can be realized in a low-threshold media way through the digital's participatory dimension. Thus, individuals are brought together through "web- and app-based platforms to bring together buyers and sellers (as well as people who are interested in non-commercial exchange) of goods, orders, and services" (Hill 2017, p. 36). The digital platforms provided enable private individuals to provide services and potential customers to use these services. These digital platforms bring the concept of the sharing economy into the digital age:

> In the following, we first summarize under the term sharing economy all offers that concern the temporary exclusive use or the sharing of a durable good. The use can also be associated with a service (for example, cleaning the apartment after a rental through Airbnb or driving in the case of a booking with Uber). Temporary in this context means no transfer of ownership, but only a rental, that is, the good reverts to the owner after the rental (Peitz and Schwalbe 2016, p. 4).

The characteristic of temporary use corresponds to the ephemeral and participatory structure of the digital. The participatory connotation gives the sharing economy a supposed moral grounding: "The idea of the sharing economy sounds super – environmentally correct, non-partisan, anti-individualist, and all wrapped up in the cuddly warm vocabulary of 'sharing'" (Hill 2017, p. 37). A paradigmatic example of such a sharing model is Airbnb, where private rooms and apartments are brokered to travelers. Airbnb, an acronym for 'Air and Breakfast' or 'Air Mattress and Breakfast', was founded in 2008. With its business model, Airbnb competes with established hotels and drew criticism for, among other things, misappropriation of living space. Simultaneously, alternative business culture is becoming visible at Airbnb, which is located beyond established business models. This alternative business culture promises customers a cost advantage and at the same time gives the company or the start-ups a rebellious semantics:[2] "The CEOs of many of these companies are brilliant at touting their platform as the greatest potential advantage for their customers. The platforms benefit from an aura that combines convenience with a touch of revolution. Convenience *as* revolution" (Hill 2017, p. 37, e.i.o.). From this perspective, the sharing model of digital start-ups reads as updating a traditional narrative of rebellion against established business models. Start-ups such as Airbnb evade regulations to establish companies such as hotels are subject (see Peitz and Schwalbe 2016, p. 22). What is lost sight of is that the digital sharing economy's alternative business culture is shaped by an ethic of profit maximization and efficiency. For example, while Airbnb has created "a popular alternative to hotels that is both cheaper and more convenient for travelers and allows people to earn extra money by renting out vacant rooms" (Hill 2017, p. 12). However, Airbnb is now "infiltrated by professional real estate firms [who] know they can double their income by pushing tenants out and keeping entire houses free for tourists" (Hill 2017, p. 12). As practiced by Uber or Airbnb, Stalder sees a sharing economy as "undermining the idea of communal use and provision of resources" (Stalder 2016, p. 278). In this context, "terms like 'open' or 'sharing' only serve to give hypercapitalist structures a [...] positive claim" (ibid.).

Despite the neoliberal practice, a metaphor of rebellion is established that weaves through the start-up discourse. The semantics of 'Californian ideology' or Silicon Valley, which is characterized by geographical proximity to San Francisco, also plays a role. This central location of the IT and high-tech world from a discursive perspective represents the southern part of the San Francisco Bay Area, which forms the metropolitan region around San Francisco and San José. San Francisco, in turn, has been "[v]irtually an epicenter of disruption and revolution, whether it is hippie flower power or beatniks, anti-war programs and the free speech movement, LSD, free love, or the women's movement and same-sex marriage" (Hill 2017, p. 8f.). The creative, innovative spirit of start-ups and their unconventional, supposedly anti-establishment innovation practices seem to fit seamlessly into this paradigmatic set of alternative cultural practices. The discursive transformation

---

[2] Hill proposes to make the neoliberal economization of the sharing economy in the digital age usable again in the sense of an alternative culture. He contrasts the concept of platform capitalism with the concept of platform cooperativism. Platform cooperativism "offers the concept of new ownership models for the Internet [...] Why can't a sharing service like Airbnb be owned and managed by the hosts or the landlords who, after all, contribute the most to the value of the business? The internet reduces to some extent the costs of building new online services and platforms, so they could well be democratically controlled and managed within reasonable cost limits" (Hill 2017, p. 219).

process that leads to the new spirit of capitalism and unfolds in the digital age in the figure of the start-ups finds here a symbolic, geographically bound localization.

### 3.2.1   'Fired by Algorithm': Start-Ups as Drivers of Precarization

Start-ups such as Airbnb and Uber, which transfer the sharing economy's approach into neoliberal contexts and carry it into the digital age, enable the individual to act as an entrepreneurial self via the low-threshold world of Web 2.0. Through the Internet, services can be offered on digital platforms. The participatory dimension of the digital cultures provides the basis of neoliberal freedom, allowing everyone to be "the master of their skill" (Hill 2017, p. 38). Behind the euphemized neoliberal flexibility and 'liberation' of service providers, a "platform capitalism" (Hill 2017, p. 36) emerges. Digital media are deployed in the interests of rationally efficient cost reduction. Businesses "powered by new technologies make it easier than ever to hire and fire employees. This further dismantles the social components of labor markets" (Hill 2017, p. 36). Digital platforms offer the possibility space for private individuals to become professional entrepreneurs, constituting a form of radical precarization that unfolds on neoliberal economics. Thus, "the boardrooms of many of these U.S. companies follow the philosophy of extreme 'economic liberalism,' eschewing all government and favoring labor whose use they can turn on and off like light bulbs" (Hill 2017, p. 14). One way in which the discursive ideal of the entrepreneurial self-manifests itself in the labor market is in the form of the subcontractor. US start-ups, for example, employ "a massive amount of subcontractors, freelancers, temporary workers, or so-called solo self-employed (self-employed without employees), whom they can hire and fire at will" (Hill 2017, p. 14). The flexibilization of employment relationships corresponds to a free market's imperatives in the sense of neoliberal concepts. This leads to a culture of precarization or a culture of stable instability in the context of employment relationships. In terms of a neoliberal "hyper-efficiency, the 'free' areas of a working day are now curtailed" (Hill 2017, p. 48). The "value of a worker" (ibid.) is thereby reduced "to the precise number of minutes" needed to, for example, "paint a garden table, design a website [...] There are no fixed annual or monthly salaries. It's like a top footballer like Miroslav Klose only getting paid for the goals he scores" (Hill 2017, p. 48). This reduction dissolves a 'partnership employer-employee relationship' in favor of cost-benefit calculations (see Hill 2017, p. 14). The separation of individuals in terms of the isolating logic underlying the entrepreneurial self's narrative casts each individual back upon themselves. As the partnership relationship between employee and employer erodes, so does social security that can accompany employment relationships. Thus

> [v]irtual digital companies [...] pay low wages, offer no social security or health insurance, and are not committed to a partnership relationship. They can simply jettison workers by excluding them from the digital platform without warning or notice: *fired by the algorithm* (Hill 2017, p. 14, e.i.o.).

In the pursuit of "hyper-efficiency" (Hill 2017, p. 48), in terms of panoptic monitoring strategies, the employee is measured, evaluated, and, if necessary, dismissed for poor performance. The "work performance [is] regularly monitored, analyzed and judged by the company" (Hill 2017, p. 49). Thus, in the digital age, a social structure is established that can be analytically worked through with the concept of Deleuze's control society.

### 3.2.2   The Cybernetic Capitalism of the Control Society

Monitoring constitutes a control regime in which the 'customers' have an evaluative-disciplinary effect on the employees. Thus, tracking on the part of companies is flanked by tracking on the part of customers. The buzzword here is "customer satisfaction, as consumers also rate performance with one to five points or stars, all logged in the smartphone, which now resembles an algorithmic overseer rather than an instrument of liberation" (Hill 2017, p. 49). Cybernetic capitalism is establishing itself (see Schaupp 2016), in the course of which feedback as a 'quality measurement' is advancing to become the powerful instrument of control: Schaupp points out that "cybernetic capitalism can be characterized as a fusion of neoliberalism and cybernetics" (Schaupp 2016, p. 83).

Cybernetic capitalism is "characterized above all by the fact that information processing, control, and capital accumulation fall into one" (Schaupp 2016, p. 83). Workers receive digital feedback, which can also be understood in terms of interpellations. Based on the feedback, workers can realign their actions, to optimize their behavior and thus themselves or their entrepreneurial selves. At the same time, feedback also allows for employee control. However, "[t]he same technologies [...] are also used in the course of monitoring and rationalization measures, both on a business level and a macroeconomic level" (Schaupp 2016, p. 83): "If the rating is too low, you are automatically excluded from the platform. *Fired by algorithm*" (Hill 2017, p. 49, ed.). In this logic of cybernetic capitalism, the customer is also included in controlling feedback loops:

In the rating system used by Uber, both the driver and the passenger give a mutual rating of their behavior (cleanliness of the vehicle, driving behavior, friendliness, punctuality, etc.) by awarding between one and five stars (maximum) to the other side of the market. If the driver's average rating falls below a certain level, he or she must expect not to be considered a driver in the future (Peitz and Schwalbe 2016, p. 11).

The digital platform capitalism of start-ups is characterized by a digital peer-to-peer panopticism. The feedback is the tool of mutual control. The digital platform capitalism of start-ups demands a 'flexibility' that is discursively clothed in neoliberal semantics of freedom. The control function via feedback, which takes the form of cybernetically controlled quality management, can be defined as a general feature of the social networking universe: Via Likes, Thumbs, etc., people evaluate each other and thus evoke a polyphonic and directional dynamic. Bröckling sees in these feedback processes a control dynamic that governs the 'self-regulation' or self-control of the entrepreneurial self: "The individual appears as an information-processing system that flexibly adapts itself to the expectations of its environment if only it is regularly fed with differentiated feedback" (Bröckling 2013, p. 239). Feedback-supported self-control can be analyzed as a form of permanent interpellation, which is generated via feedback. Thus, "feedback loops are installed that signal deviation from the norm to the individual but place the necessary adaptive efforts under his or her own responsibility" (ibid.). Self-control, which is enabled by interpellative invocations, corresponds to the efficiency logic of neoliberal approaches and is enabled by technologies of domination. These technologies of domination lead away from the disciplinary society towards the control society: "In short, entrepreneurial selves are not fabricated with the strategies of surveillance and punishment, but by activating the potentials of self-control" (Bröckling 2013, p. 61). Unlike the disciplinary society still described by Foucault, the control society does not confine individuals under coercion (see Deleuze 2005, p. 7). Instead, in terms of discursive control, actors are 'motivated' to self-optimize by the argumentation of constraint (the rational market demands). Deleuze analyses a cross-field penetration of market logic that leads actors to self-optimize as products. In doing so, Deleuze refers to "forms of continuous control and the impact of permanent education on schools, accordingly the abandonment of all research at the university, introducing the 'enterprise' at all levels of education and training" (Deleuze 2005, p. 13). Ehrenberg (2015) speaks of "personal initiative, submission to the norms of performance: personal initiative is necessary for the individual to remain socially acceptable" (Ehrenberg 2015, p. 299). In these processes, control manifests itself in the actors' behavior and actions and not necessarily in their intentionality. A characteristic of the control society is that actors keep themselves in constant motion while experiencing stable instability. Through this ephemeral structure of the control society, precarity emerges as a form of life and experience:

> In disciplinary societies, one never stopped starting (from school to barracks, from barracks to factory). In contrast, in societies of control, one never finished with anything: Company, further education, service are metastable and coexisting states of the same modulation (Deleuze 2005, p. 9).

The change from a disciplinary society to a society of control is also medial. Thus, the "old societies of sovereignty [...] dealt with simple machines: Levers, pulleys, clocks; the recent disciplinary societies were equipped with energetic machines" (Deleuze 2005, p. 11). "Societies of control," on the other hand, "operate with machines of the third kind, information machines, and computers" (ibid.). To illustrate this, Deleuze draws on a future scenario designed by Guattari, which prefigures control effects that are already partially a reality within the SNS universe:

> Félix Guattari imagined a city in which everyone could leave his apartment, his street, his neighborhood thanks to his electronic (divisional) card, through which this or that barrier opens; but the card could also be invalid on a certain day or *for* certain hours; what counts is not the barrier, but the computer, which registers the – permitted or unpermitted position of each individual and carries out a universal modulation (Deleuze 2005 p. 12f.).

In the control society, spaces are stretched out, and structures are established that pre-figure the behavior of the actors. Digital media can help to fix the individual and track their movements in the spaces – an aspect that already appears as a reality in the context of digital self-measurement. Authoritarian structures of domination are reduced to the entrepreneur's figure or the *"corporate regime"* (Deleuze 2005, p. 13, e.i.o.), which has parallels with Bröckling's metaphor of the entrepreneurial self. This is defined by the individual's hope to gain and maintain agency through entrepreneurial virtues. Simultaneously, with the entrepreneur's enthronement as a metaphor of domination in control society, the boundaries between the state and the corporation are leveled. Thus "the factory has given way to the enterprise" (Deleuze 2005, p. 260):

> Family, school, army, factory are no longer different analogous milieus converging on one owner, state, or private power but are ciphered deformable and transformable figures of the same company, which only knows managing directors. Even art has left the closed milieus and enters the open circuits of the bank. The conquest of the market happens by seizing control and no longer by disciplining (Deleuze 2005, p. 12).

Control can be achieved through quantification. In a society of control, individuals have become *"divisible"*"(Deleuze 2005, p. 10, e.i.o.). Individuals become elements that can be divided and related – individuals or their actions are recorded as data. These data form the basis for a "numerical language of control" consisting of "ciphers." These ciphers or numerals generate information through "masses, samples, data, markets or 'banks'" (Deleuze 2005, p. 10). This analysis by Deleuze reads like an analytical perspectivization of the phenomenon of Big Data:

> Big Data results not only from online data, customer data, and user-generated content as 'unstructured data' (the content of emails and SMS messages, etc.) but also arises equally at the interfaces of automated data processing. In this sense, applied network research speaks of transactional user data generated by web tracking, mobile phone monitoring, or sensor capture (Reichert 2014, p. 10).

Here, contours of the digital control sketched by Guattari and Deleuze appear, which simultaneously construct data and, in the course of evaluating this data, ciphers. Thus, the handling of extensive unstructured data, which are generated via social media, enables the construction of knowledge. "From a systematic analysis of Big Data, forecasters expect more efficient corporate management in the statistical measurement of demand and sales markets, individualized service offerings, and better social governance" (Reichert 2014, p. 10). Through digital data, supra-individual knowledge in the sense of biopower is generated. At the same time, in the sense of cybernetic capitalism, individual data act as feedback that enables neoliberal self-optimization: the feedback of cybernetic capitalism produces control mechanisms that lead to

- "the transparency society acquiring a structural proximity to the surveillance society" (Han 2013, p. 91) and
- taking on contours of a control society.

In the digital age, the control society is characterized by an inseparable dialectic of digital governmentality and economic exploitation of the digitally captured individual:

> The Internet and communications corporations seem [...] not only to show an astonishing willingness to cooperate but are themselves significant drivers of an enormous accumulation of private data and its evaluation for economic purposes. Big Data is a commodity of the globalized media economy, and the digital society is no longer free to decide whether it wants to escape this availability (Kammerl 2017, p. 44).

With (self-)control through digital data, distrust manifests itself subtextually: In the course of the social transformations evoked by digitalization, the digitally readily available information that forms the basis for feedback metonymically effects a loss of trust: "Where information is very easy and quick to obtain, the social system switches from a trust to control and transparency logic" (Han 2013, p. 91). This loss of trust also follows an "efficiency logic" (ibid) and can be understood as a governmentality practice of neoliberal power structures. Feedback becomes an objective instance of knowledge and undermines the autoreferentiality founded by Descartes as the basis of valid knowledge. Reflexive self-assurance, phenomenological intentionality, hermeneutic understanding, and pre-reflexive intention are replaced by the objectivity of feedback coming from outside. Self-assessment is made dependent on feedback information, leading to an alienation of one's epistemological competencies and self-awareness. This alienation from one's autoreferential forms of cognition, justified by feedback, conditions a distrust of one's own self-perception/world-perception. The feedback becomes an objectifying corrective to the autoreferential self-world perception.

The argument developed can be summarized as follows: The ephemeral as a structural feature of the digital and the decentralized polyphonic poly-directionality of the Internet can be narrated in the context of neoliberal discourses as a medial realization of the market's demands for freedom. In the entrepreneurial self's metaphor, this concept of freedom totalizes the postmodern moment of epistemological self-responsibility, suspending aspects such as social equality. In the process, neoliberalism establishes itself as a metanarrative in the digital age, finding expression in the discursive enactment of "start-up fever" (Hill 2017, p. 8). The flexibility and concomitant state of precariousness experienced by the neoliberal individual are flanked by control of performance. This gives rise to the paradox of a normative space of possibilities: in terms of the entrepreneurial self, the individual has the opportunity to develop in new project-based challenges. In doing so, new work contexts and communication processes can always be learned. Ideally, this leads to a constant expansion of competencies and successive 'self-development through work.' This requires flexibility and the ability to endure precariousness. At the same time, self-exploitation is monitored in the sense of neoliberal efficiency logic. From this perspective, the neoliberal individual is embedded in a normative space of possibility in which prefigured efficiency parameters demand a flexible self-exploitation. Freedom in the sense of an epistemic self-location is replaced by freedom of efficient self-utilization: the subject does not fulfill itself in the freedom of reflexive self-location. Instead, socially normative spaces of possibility constitute individuals' self-perception and emotional self-experience through interpellative subjectivation processes. Discursively, this normative possibility space is "legitimated by a charismatic ideology which ascribes to the person, to his natural gifts or merits, the entire responsibility for his social destiny" (Bourdieu 1983, p. 606). The normative space of possibility and its subjectivizing dynamics are significant features of the social network sites (SNS) universe. Here, the potentials of freedom of the Internet find a neoliberal framing. With this neoliberal which forms of control society in are also constituted. While the Internet's emancipation potential can be seen in the 'becoming liquid' of the individual, the SNS universe fixes the individual through (neoliberal) narratives in digital space.

## 3.3   The SNS Universe

The use of digital media as social media is increasingly inscribing itself into the lifeworld, shaping our everyday lives and profoundly restructuring cultural practices. By being embedded in the social web of social networking sites (SNS), the individual is constituted as a social coordinate in the digital age. Through communication platforms such as Facebook, Google+ and Instagram, Twitter, Snapchat, and LinkedIn, users constitute their "personal cognitive horizion [...] in countless streams, updates, and timelines" (Stalder 2016, p. 139). In the course of this constitution process, "the most important resource is the attention of others, their feedback, and the resulting mutual recognition" (ibid.). The monadic starting point of this social dynamic is the creative individual. This (neoliberal) individual produces data by being embedded in social networks, thus exhibiting the feature of participatory content generation that characterizes digital cultures. Simultaneously, the user consumes the data of other users, and comments on it. In the form of his 'participatory consumption' of content, the users generate content by commenting on it. In this oscillating positioning between 'producer' and 'consumer' of content, the actor in the SNS universe becomes a 'prosumer' (see Toffler 1980):

> The prosumer can take different forms in the Internet economy. The most obvious is the gratuitous production of content for commercial software providers. This is the case, for example, with most commercial social media platforms, in which photos and other data become the property of the operating company in the course of self-representation and evaluation (Schaupp 2016, p. 79).

In this social dynamic, the user "becomes a product himself" (ibid.) when he consumes advertising. Consumption and production merge, with the valorization of data being the driving business model of SNS:

In the front end, the website's user-visible surface, information is displayed, or goods are sold. In the back end, information about the users is collected at the same time. This information is resold in aggregated form or the form of personalized data as goods. Thus, a new level of capital accumulation is added. In addition to the advertising that users consume on the website itself, a second value creation process occurs through data collection (Schaupp 2016, p. 80).

The social networking sites form a universe in which users create profiles, share data, evaluate each other. This SNS universe can be understood as an expanding field of neoliberal ventures, the growth of which is shown in Fig. 3.1. In the SNS universe, users are increasingly embedded in a normative space of possibility,

- In which they narrate themselves as actualizations of the entrepreneurial self and,
- evaluate each other via feedback.

The social networking sites constitute 'a totality of the Internet' or form a 'supposed totality of the Internet': it is possible to move around the Internet exclusively via SNS and to be accompanied continuously or 'guided' by SNS services (Google+, Google Maps, Google Calendar, Google Notes, etc.). From this perspective, the network of SNS forms a (potential) universe/a whole, which appropriates the users via different applications. This self-contained dimension of the Internet takes place in times of Web 2.0: for example, Google exists since 1998, Facebook since 2004 and YouTube since 2005. The poly-directional and polyphonic possibilities of Web 2.0 are constituted in the context of commercialization of the Internet. The "new social mass media, such as Facebook, Twitter, LinkedIn, Instagram, WhatsApp or most other commercial services developed after 2000" (Stalder 2016, p. 214) are based on "closed standards controlled by network operators" (Stalder 2016, p. 214). These standards form the boundaries of the expanding SNS universe and "prevent users from communicating across the boundaries defined by the providers. Through Facebook, one can only get in touch with other users of the platform; those who give up their accounts also end their Facebook friendships" (Stalder 2016, p. 214f.). The expansion of the SNS universe results from the network effect – "a network becomes more useful and attractive the more people it connects" (Stalder 2016, p. 231) – which "becomes a *monopoly effect*" (ibid., e.i.o.):

The whole network can consist of only one supplier. This connection between network and monopoly effect is not compelling but contrived. The closed standards make it impossible to change providers without losing access to the whole network

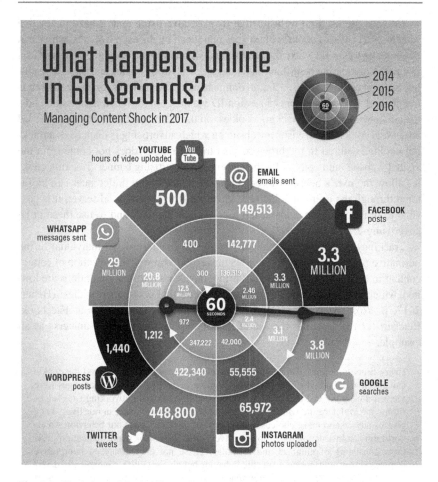

**Fig. 3.1** The infographic visualizes what happens in a minute in the social networking universe. The reference to the annual numbers shows the expanding size of the SNS universe (Source: http://www.smartinsights.com/internet-marketing-statistics/happens-online-60-seconds, last accessed: February 19, 2017)

and thus also to the community formations that were created on its basis (Stalder 2016, p. 231f.).

The closure of movement spaces or the restriction of freedom of movement can also be seen in so-called tracking cookies.

Cookies are used on websites that display advertising and/or have interactive functions. The use of cookies often goes unnoticed by the user. Cookies, which can be described as a tiny text file, enable the webserver to recognize a visitor to a website. Cookies can be differentiated between session cookies and tracking cookies. Session cookies are used a.o. in online banking and exist as long as a session is maintained. The server stores a session ID via the cookie. When the session ends, the cookie is dissolved. Tracking cookies, on the other hand, are used by so-called ad servers. When a user visits a website on which advertising is displayed, a tracking cookie is stored in the browser. This tracker mostly does not come from the visited website but is placed by the ad server's advertising banner. Tracking cookies allow the user's behavior to be analyzed: Interests and preferences can be filtered through the user data obtained. Via a tracking cookie or ad server, it is possible to draw conclusions about the user or their interests and to use these in an advertising-effective manner.

Tracking cookies paradigmatically show the fixation of the individual in the normative possibility space of the SNS universe: "Through the selection of certain content in the form of sorting and filtering, the user's information options are limited, which can be illustrated by the personalization of newspaper articles" (Hebert 2017, p. 76). The freedom of movement is thus prefigured. About Facebook, Reichert (2013) describes the use of tracking cookies in the SNS universe as an example:

> Every time an online visitor returns to Facebook, the server that created the cookie can check and read what was previously written to the file, such as what pages were viewed during the last user session. On the one hand, social networking sites' algorithms try to determine what is relevant and what is less relevant in our lives. They give us advice and make choices and actions for us to influence our behavior. An algorithm is an instruction for action to solve a task. Its basis is mathematical propositions consisting of numbers and formulas. These not only create an interpretive framework for interpreting a specific behavior but also establish a technically determined frame of reference and preference into which we design our lives (Reichert 2013, p. 64).

Reichert's description can be read as a process of constitution of the space of possibility spanned by SNS. Within this space, users' movements can be totally recorded. "Every click I make is stored. Every step I take is traced. We leave digital traces everywhere. Our digital lives map exactly onto the web" (Han 2013, p. 92). Han sees in this the "possibility of a total logging of life" (ibid.), which would lead to the completion of the transparency society (see ibid.). Privacy and intimacy are abandoned or give way to the "need [...] to display them shamelessly", which leads to "freedom and control becoming indistinguishable" (Han 2013, p. 93).

The effective power of the SNS universe is paradigmatically demonstrated when no other space beyond this world of neoliberal self-dramatization seems conceivable: "While users are advised to be data-savvy, the empowerment to create and shape their own social spaces on the net is largely ignored, because this would initially mean turning away from social services such as Facebook & Co. that are attractive to users" (Biermann and Verständig 2017, p. 8). The ambivalence between freedom and control is also evident in the interlocking of neoliberal self-thematization in the digital space of the SNS universe and the SNS user's valorization. Thus, the "subjectivities portrayed in social networks [...] can be seen as expressions of identities and subjectivities" (Hebert 2017, p. 74). On the other hand, these profiles represent "no more than a collection of bits" (ibid) that "objectifies users into 'information objects' [...] for monetary gain" (Hebert 2017, p. 75). Neoliberal freedom becomes a metanarrative of the control society of the digital age. Control and profit generation appear as inextricably intertwined.

In the SNS-universe, neoliberal flexibility of the individual is realized through control or fixation of the individual using a sociometric registration via social software profiles. The openness of the Internet, which is also a constitutive topos of the semantics of freedom in the cyberspace discourse, is suspended.

As in the culture of platform capitalism of the sharing economy, 'flexible self-location' and panoptic (self-)control can be identified as significant features of the SNS universe. These features constitute the normative space of possibility for the unfolding of the entrepreneurial self in the SNS universe's digital sphere. Social networking sites such as Facebook, Tinder, Instagram, and Google+ reveal ambivalent semiotics: on the one hand, they 'fix' the individual and unfold interpellative dynamics; on the other hand, their poly-directional and polyphonic structure enables a participative culture. This participative culture presupposes the individual as a monad who gains access to the respective community through a profile. The communities span the normative space of possibility that prefigures the individual's possibilities of movement: "In view of a digitally shaped world of life, it becomes apparent that the spaces of democratic and dialogical negotiation of information are dwindling. The large corporations form platforms in which they define the rules and norms and (pre)structure communication" (Krückel 2017, p. 54). These prefigurations of movements through digitally expanded normative spaces of possibility fix the individual in terms of the control society. It is precisely in this fixation of the individual that is a difference from the semantics of freedom in cyberspace: the semantics of freedom in cyberspace evoke collaborative forms of knowledge generation. In the SNS universe, the individual forms the starting point. Already in the process of registering with a social networking site such as Facebook, the individual is subjectified through interpellative self-identification. In the act of registration, the individual locates him or herself within socially conventionalized parameters that define the individual (age, profession, gender, etc.). In this way, the individual integrates himself into social matrices of the material-physical world and thus performatively reproduces its hierarchies and dependencies:

> The registration process has remained very simple but once registered, users are continuously faced with prompts to provide personal information which enables them to be categorised [...] Power is made manifest on Facebook in various ways: the constant prompts urging users into self-revelation; the constant threat of exclusion if users do not provide access to personal information; and the lack of control users' have over their own information and content (Buchanan 2011, S. 275f.).

The individual is confronted with preconfigured narrative topoi and must use them if he or she wants to 'socialize' based on SNS.[3] Accordingly, Faucher (2013) elaborates that SNS represent "obfuscated prison-houses" (Faucher 2013, p. 9). These "guide and direct human behavior in their environments, making use of several prompts and cues that constrain choice under the illusion of freedom" (ibid). An example of such an interpellative form of self-narration is Facebook's so-called Chronic feature. The plurality of identities of the postmodern, 'fluid' subject is substituted by a linear self-narrative, which the Chronic-feature demands. Rather than generating different identities in diverse contexts, Facebook constitutes one identity. This singular identity of the user is confronted with the imperative to act socially. In the SNS universe, social action is defined by evaluative interaction: there is a need to 'like' and upload posts, which can be liked. Economic calculations guide the recognition procedure that enables entry into this world of evaluative interactions:

> Almost all commercial social mass media are financed by advertising. Facebook, Google, and Twitter generated ninety percent or more of their revenues from advertising in 2014. Accordingly, these companies need to find out as much as possible about their users to optimize access to them for advertising customers and be able to sell them (Stalder 2016, p. 219).

In the SNS universe, the individual must be permanently social to assert itself as a monad in the SNS universe. If an individual does not act accordingly, the individual dies the symbolic death in the SNS universe. Evaluative interaction appears

---

[3] Reichert (2013) points out that "under all circumstances there is a multitude of tactical possibilities" (Reichert 2013, p. 60) to subvert the "form-immanent[e] dictate" (ibid.) of the registration process. At this point, the analysis of the interpellative dynamics is in the foreground, so that possible subversion strategies are not included in the analytical focus here.

necessary to keep the prosumer actively moving, thus ensuring the dynamics of cybernetic capitalism. Movement is consequently a feature shared by postmodern epistemology and the neoliberal entrepreneurial self. However, while postmodern epistemology aims at dialogical agility, the entrepreneurial self's movement is a movement of self-optimization in course of the competition on the neoliberal market. Social existence in the SNS universe is associated with normative pressures. Evaluation is always present. Paradigmatic for this logic for the assessment in the SNS universe is the so-called Facebook thumb. From a semiotic perspective, the Facebook Thumb can be read as a performance evaluation for successful self-narration: likes can measure the impact of self-narrations. The Facebook thumb or Facebook thump-up (and variations on other SNS) represent a "digitized gesture signaling approval, approbation, agreement, praise or even on occasion a reminder to the receiver of the sender's existence" (Faucher 2013, p. 1). Evaluations are an integrative element of the SNS universe that relates users to each other. This social dynamic of evaluative interaction challenges academia as a research field (for an overview of empirical research on Facebook, see Nadkarni and Hofmann 2012). Bak and Kessler (2012) elaborate on conformity effects in Facebook use by examining Like behavior in relation to a posted image. They varied how many Facebook likes an image was presented with. It was found that intensive Facebook users, in particular, liked a picture better if it had already been positively labeled by likes (see Bak and Kessler 2012, p. 23). Against the background of Foucault's model of the panopticon, it can be argued that the communicative structure of SNS unfolds an intersubjective surveillance effect: "Every social media user can be equally observer and observed, controller and controlled" (Mitrou et al. 2014, p. 12). Andrejevic (2005) describes this phenomenon with the term peer-to-peer monitoring: "[P]eer to peer monitoring understood the use of surveillance tools by individuals, rather than by agents of institutions public or private, to keep track of one another, covers (but is not limited to) three main categories: romantic interests, family, and friends or acquaintances" (Andrejevic 2005, p. 488). SNS performatively reproduce panoptic power structures and an economic logic in which an increase in likes can be interpreted in terms of positive self-presentation. This economic logic also allows companies to present themselves through Facebook pages. In this way, they move closer to the private sphere and allow themselves to be included in the circle of friends. The spheres between the private and the public/commercial are dissolving. The blurring of the boundary between these spheres can be understood through the results of a study conducted in 2012 by Harris Interactive on behalf of Careerbuilder, an online job portal. The study surveyed 2303 hiring managers regarding the importance of social media in the hiring process. The recruiters said that to check the suitability of an applicant, they also look at the applicant's social media activities. The main source was Facebook with 65% followed by LinkedIn with 63%. The reviewing view of HR managers is directed at the fit between the applicant's self-narrative and the company's self-image:

- *When asked why they use social networks to conduct background research, hiring managers stated the following:*
- *To see if the candidate presents himself/herself professionally – 65%*
- *To see if the candidate is a good fit for the company culture – 51%*
- *To learn more about the candidate's qualifications – 45%*
- *To see if the candidate is well-rounded – 35%*
- *To look for reasons not to hire the candidate – 12% (Careerbuilder 2012, Section 7, e.i.O).*

The view of HR managers has a normative disciplining effect. For example, 34% of recruiters who have already used SNS to review a candidate stated that they had found the information in the SNS narratives of the candidates that prompted them not to hire them. The reasons cited range from uploading inappropriate content to candidates lying about their qualifications:

- *Candidate posted provocative/inappropriate photos/info – 49%*
- *There was info about candidate drinking or using drugs – 45%*
- *Candidate had poor communication skills – 35%*
- *Candidate bad-mouthed previous employer – 33%*
- *Candidate made discriminatory comments related to race, gender, religion, etc. – 28%*
- *Candidate lied about qualifications – 22% (Careerbuilder 2012, Section 10, e.i.o.).*

29% of recruiters stated that they found information on SNSs that positively influenced them when hiring an applicant. The reasons given for the positive impression conveyed via SNS range from intuitive assessments to positive feedback or that other users have posted positive things about the applicant:

- *Good feel for candidate's personality – 58%*
- *Conveyed a professional image – 55%*
- *Background information supported professional qualifications – 54%*
- *Well-rounded, showed a wide range of interests – 51%*
- *Great communication skills – 49%*
- *Candidate was creative – 44%*

• *Other people posted great references about the candidate – 34% (*Careerbuilder 2012, Section 12, e.i.o.).

The normative view of HR managers is directed towards private life, assessed under the parameters of an applicant's employability. Willey et al. (2012) work out that the "utilization of social network sites for applicant screening will continue" (Willey et al. 2012, p. 307; see also Chiang and Suen 2015). The fact that such a view of SNS by HR managers is also affirmatively accepted by applicants can be seen in the results of a study by Martensen et al. (2011), in which "the impact of social networking sites on the employer-employee relationship" was investigated. One finding of the study is that "members of SNS do believe that (potential) employers carry out research on the Internet and that users behave accordingly" (Martensen et al. 2011, p. 252). Among other things, SNS users were asked whether they would use the Internet as a platform for a self-narrative directed at potential employers: "[T]he 228 respondents (60.7%) agreed with the following statement: *The Internet enables me to present myself the way I want to* (n=376, $\mu$=3.59, $\sigma$=0.78)" (Martensen et al. 2011, p. 250, e.i.o.).

In summary, it can be stated that the SNS universe enables self-marketing logic within the private sphere. It exists a fit between private lifestyle, which is performed on SNS, and professional requirements. SNS evoke panoptic visibility of the individual: the private becomes public. Metonymically, the focus on the individual or the 'assignable self' is also evident in the naming of networks: the addressing and simultaneous fixation of the individual in the naming of SNS can be exemplified by the video portal YouTube. "The first part of the term contains a general imperative of participation ('You'), and the second part of the compound word 'Tube' establishes a colloquial reference to the television" (Reichert 2013, p. 145). In the sense of the SNS universe's participatory culture, the user is addressed as a content-producing individual who generates the broadcast content himself. This is also reflected in the YouTube slogan 'broadcast yourself.' This marks the replacement of the television as mass medium by Web 2.0 tools, which can be understood as the digital age's mass medium. At the same time, a self-narration is interpellatively demanded, which constitutes the SNS universe as a forum for the narration of the entrepreneurial self. YouTube enables an "infrastructural empowerment" (Reichert 2013, p. 147) that is built on "standardized and normative specifications" (ibid.) and reproduces a neoliberal market logic. The drawing of attention becomes the parameter that structures competition between entrepreneurial selves which uploading videos to YouTube. Thus, "the rating and evaluation systems used on YouTube [...] stimulate quasi-markets and competitive situations that apply equally to all videos" (Reichert 2013, p. 147):

Videos that want to escape the structural ignorance of 'zero comments' [...] and generate attention above the threshold of non-perceptibility must also survive the various rating, voting, and polling procedures that YouTube offers to legitimize its web-based content. Internet videos on YouTube are organized in a network and integrated into dynamic social negotiation processes that ensure permanently changing framings and contextualizations using decentralized feedback and control procedures. Accordingly, every video content is under pressure to be perceived since only the contribution that is ratified as legitimate in the context of the YouTube standardized grading games can prevail (Reichert 2013, p. 147).

Cybernetic capitalism's logic is updated by "one of the most influential media phenomena of the present" (Reichert 2013, p. 144). It is precisely the publicity of the private that in represents a business model used by so-called YouTubers: (German) YouTube stars such as Julian Claßen (Julienco) and Bianca Heinicke (BibisBeautyPalace) narrate themselves as a couple on YouTube and thus generate a role model for relationship. YouTube clips allow YouTubers to share their private lives with the public and enter into a para-social relationship with their viewers. Since both have several million followers and their videos are viewed several million times, they can earn money through proportional revenues that result from the advertising placed before and after their video clips.

Within this context, surveillance is not a compulsory act but a cultural offer that is disseminated via SNS. In the process, a lifestyle is demonstrated that interpellatively reinforces role models. For example, when 'Bibi' gives so-called 'beauty tips' or Shirin David sells her perfume at a drugstore chain and promotes it via YouTube clips. Shirin David produces clips such as 'One day five looks'.[4] The public gaze is directed towards the private. At the same time, the private becomes public and a business model. The boundaries between private and public blur and produce an entrepreneurial self in the digital age. In the process, the spheres between the digital and the material-physical world also blur: the digitally based self-narrative via SNS is performed via videos and images from the material-physical world.

The private display becomes a business model, which was discursively prefigured by television formats such as Big Brother at the beginning of the 2000s via television – the mass medium of the electronic age. In the process, Web 2.0 media's dialogic potential is overformed by a non-democratic form of communication: You Tube stars are the television stars of the present. This can be seen metonymically when Shirin David becomes part of the jury of the television casting show 'Germany seeks the superstar' ('Deutschland sucht den Superstar'/DSDS) due to her fame as

---

[4] See https://www.youtube.com/watch?v=EEgE2t8x260, last accessed April 23, 2017.

a YouTuber (on the relationship between neoliberal lifestyles and casting shows, see Kergel 2016). YouTube stars act as role models by broadcasting their content to a mass of recipients. The sender-receiver principle is thus established in the world of Web 2.0. This closing of the participation space is reinforced by the business model underlying so-called multimedia channels such as Mediakraft: These channels 'support' individual YouTubers or specifically build them up into stars by discursively staging the YouTubers as a product. This commercialization is euphemized as 'support' for professionalization: The productions of YouTube clips are to be made of higher quality to generate more clicks for more advertising revenue. Semiotically decoded, this 'assistance' means installing the unidirectional start-up cult in the SNS universe.

With the establishment of the mobile Internet, neoliberal (self-)narrations reach out of cyberspace's digital space and inscribe themselves in the material-physical world. Whereas Turkle still saw cyberspace as a postmodern space of freedom, the SNS universe becomes a neoliberal space of control. The users' controlling measurement takes place simultaneously in the virtual and the material-physical world, thus extending the SNS universe to the material-physical world and becoming total. The dichotomization between 'virtual' and 'material-physical world' erodes from this perspective and leads to the augmented reality of the digital age's neoliberal control society. This thesis will be unfolded in the following.

## 3.4     'Internet to Go': The Mobile Internet as an Assignation of the Individual

The term mobile Internet or mobile Web refers to establishing an Internet connection via mobile digital end devices such as smartphones, tablets, or laptops. With the mobile Internet, the virtual world inscribes itself into the material-physical world. The dichotomous differentiation between the virtual world of cyberspace and the material-physical world erodes.

The mobile internet is the result of ongoing technological development: "For the first time, a commercial internet connection with mobile phones was offered in Finland in 1996, while unlimited internet browsing with mobile phones only became possible since 1999 in Japan with the mobile phone service i-mode" (Chatfield 2013, p. 74f.) In 1999, the mobile internet reached Germany when WAP (Wireless Application Protocol), a technology with 9.6 kb/s, was presented at CeBIT.[5] Since

---

[5] CeBit was the largest and most internationally representative computer expo.

then, transmission speeds have increased massively. For example, LTE (Long Term Evolution) achieves a data transmission speed of 300 mbit/s.

Evolution of technical devices flanks this development of transmission speed: IBM's 1994 Simon Personal Communicator (SPC) is the closest thing to what is now understood by a smartphone. The SPC had a touch screen, was capable of email and faxing but did not have a web browser. Among other things, because of the short battery life (about 1 h) and a high price, the SPC could not establish itself, but smartphones' development had begun. Nokia, BlackBerry, and Palm pushed the development of smartphones around the turn of the millennium. The decisive turning point came in 2007 with the introduction of the iPhone, followed by introducing the iPad in 2010. Smartphones and mobile digital devices, now supplemented by wearables – such as smartwatches and, prospectively, smartglasses – are significantly expanding the form of internet usage. The number of smartphone users in Germany increased considerably from over 6 million in 2009 to 49 million in 2016.[6] During this time, mobile data traffic in Germany multiplied from 11.47 million gigabytes to 65.41 million gigabytes. In 2015, data traffic reached 591 million gigabytes.[7] Since 2014, for the first time, more young people went online with a mobile digital device than with a desktop PC (see Feierabend et al. 2014). This rapid change led to a reconfiguration of the media structure of social reality, which also restructured digital cultures:

> The earliest stages of digital culture were closely linked to a physical infrastructure that still forms the basis of the modern Internet: a vast network of cables connecting countries and continents. Today, however, wireless connections lead a transformation that allows whole new parts of the world and all sorts of new devices to connect to the Internet. The path is from a culture of wire-connected desktops to one of countless ubiquitous wireless-connected mobile devices (Chatfield 2013, p. 72).

The infrastructural change initiated by the mobile Internet, to which the digital cultures are also subject, can be analytically processed from an epistemological or media theoretical perspective with the concept of augmented reality.

---

[6] Source: https://de.statista.com/statistik/daten/studie/198959/umfrage/anzahl-der-smartphoneusers-in-germany-since-2010/, last accessed: October 14, 2017.
[7] Source: https://de.statista.com/statistik/daten/studie/172798/umfrage/datenvolumen, last accessed Oct 14, 2017.

## 3.4.1   The Culture of the Digital as Augmented Reality

The separation between a material-physical and a virtual space does not correspond to the complexity of media use. This thesis can be elaborated based on a critical examination of the concept of media: Palm (2006) speaks of a "*conceptual chaos that the term 'medium' triggers*" (Palm 2006, p. 49, e.i.o.). This chaos, Palm continues, "could be escaped by denying it [the concept of media] any categorical meaning beyond its diffuse use for certain technical instruments of dissemination" (Palm 2006, p. 49). According to Palm, the separation between media content and media form is eroding. Media content and media form merge into one another. This thesis can be justified with reference to Kant's approach of constructivist epistemology. Instead of analyzing the medium as a "thing in itself" and in the course of this differentiating between media content and media form, a phenomenological perspective on the medium can be adopted. From this phenomenological perspective, media form and media content constitute each other. This consideration will be elaborated in the following.

In the sense of a constructivist understanding of reality, no ontological justification can be provided by the essence of things since this essence is not accessible to human cognition. This epistemological aspect, which can be read as a continuation of the autoreferential doubt, Kant grasps with the concept of the "thing in itself" (Kant 1956, B341ff., A284ff.). From the perspective of Kant's epistemology, the existence of the essence of things or a "thing in itself" represents an unverifiable hypothesis. In the process of engagement with 'the outside,' objects of cognition are constructed by the cognizing subject through their 'outside-of-the-subject-being':

> The ability (receptivity) to form ideas through the way we are affected by objects is called sensuality. Through sensuality, then, objects are given to us, and it alone supplies us with impressions; but by means of the understanding they are thought, and from it, concepts arise (Kant 1956, B33, A19).

In this context, Mersch (2006) speaks of a "paradox of the medial" (Mersch 2006, p. 224), which is characterized by a "disappearance of the medial in the appearance" (ibid.): The appearance constitutes reality as such but requires the subject to be perceived or constructed as an appearance. Through the appearance of Cyberspace's digital world, the perception of the world is extended or 'augmented' by an additional medial dimension. From this perspective, the term augmented marks a process in the course of which the perception of the world is medially expanded. From this perspective, reality becomes a term of regulation. With reference to

Kant's epistemological critique, it can be concluded that 'reality in itself' does not exist but is constituted medially in the process of perception. By adding a further medial sphere, the possibilities of constituting appearance increase or are expanded or augmented. This expansion of reality is the effect of technological developments: the carriage, the railway, the television, the telegraph, and the telephone each caused medially conditioned expansions of the perception of reality (see Hartmann 2000). In terms of research strategy, a distinction can be made between a 'technical' and a 'semiotic dimension' in the analysis of the media augmentation of reality:

> Augmented Reality is a relatively new technology that makes it possible to supplement users' current field of perception with digital media and additional information. In particular, due to developments in smartphones in recent years, it is now possible to provide mobile augmented reality applications for end-users without special hardware requirements [...] Essentially, mobile augmented reality uses the sensor technology built into smartphones, such as satellite positioning, digital compass, and gyroscope, to provide users with a selection of information on their end device (Specht et al. 2013, p. 62).

The use of digital devices is not only a technical phenomenon but also a social one. For example, social software applications such as Facebook often serve for self-articulation and thus fulfill a social function. Consequently, digitally augmented reality is also a *socially* augmented reality. This socially augmented reality is structured, among other things, by neoliberal narratives that have established themselves in the SNS universe and discursively update the model of the entrepreneurial self. Through the mobile internet, media reality is expanded. The Internet becomes an 'Internet to go' through which new cultural practices are made possible. Cultural practices established in the SNS universe that evoke the entrepreneurial self also inscribe themselves in the material-physical world through the 'internet to go.' The digital-based society of control becomes a post-digital society of control as the digital interacts with bodily (self-)disciplining to create a holistic subjectification that gives rise to the entrepreneurial self. Through this ubiquity of the digital, the digital 'disappears in appearance.' This can be exemplified by self-tracking, digitally-based self-measurement, or lifelogging.

## 3.4.2  Self-Tracking as an Expression of the Zeitgeist

Self-tracking refers to the practice of measuring one's activity via activity trackers such as fitness wristbands. In this way, steps taken, calories burned, and sleep can be recorded quantitatively. The data collected via activity trackers are sent to smart-phones or recorded via the smartphone itself:

> Mobile sports and fitness trackers are miniaturized computer applications that are mainly used for their users' physical training. They have various biometric feedback loops and develop a variety of multimedia instructions designed to get their users to engage with numerical values in specific ways. Time progression charts, audio signals, scenario charts, mean value calculations, progress, target value, and regression analyses take on action initiatives themselves by setting action goals and inviting independent responses from users (Reichert 2016, p. 192).

Generalized, self-tracking or lifelogging is about "capturing human life in real-time by the digitally recording body, behavioral, and data traces and keeping them on hand for later retrieval" (Selke 2016, p. 3).

As with the development of the Internet, the US DARPA has a central role in developing digital logging of activities. For example, the LifLog project was concerned with "comprehensively equipping soldiers of the future with sensors. The researchers' goal was to record all of the soldiers' activities from multiple perspectives to provide the operational command with a better overview" (Selke 2016, p. 2). The idea emerged to use, among other things, "a helmet-mounted, high-resolution mini-camera, plus two microphones (one for speech capture and one for ambient noise capture), a GPS location system, and accelerometers on various parts of the body and the weapon" (Selke 2016, p. 2). As a technical innovation, a semiotic/cultural dimension is also inherent in self-tracking as a social practice. In the course of establishing neoliberal narratives, the concept of self-measurement via mobile digital devices "seemingly reflects the zeitgeist perfectly" (Selke 2016, p. 5):

> According to a study by the market research company *YouGov,* 32 percent of German citizens can imagine sharing health-related data with health insurance companies to receive benefits. One in five respondents is even considering the digital measurement of their children. But most of those surveyed also have a sense of the downsides of self-measurement: 73 percent suspect that if their health deteriorates, their health insurance company would have to raise premiums if it integrated self-measurement data into the calculation models for premium rates. And as many as 81 percent believe that their data will be used for other purposes (Selke 2016, p. 5, ed.).

Selke's account highlights the ambivalence of the practice of self-tracking between the possibility of (self-)control in the precarious times of the neoliberal entrepreneurial self and the surveillance technologies of the order society of the digital age. Against the backdrop of this account and the narratives enabled by self-tracking, it is possible to state with Gugutzer (2016) that *"self-tracking is an objectification of the zeitgeist"* (Gugutzer 2016, p. 162, e.i.o.).

The Zeitgeist model was defined most effectively by Hegel and can be understood as a manifestation of ideas, views, and conscious ways of life that paradigmatically represent an epoch. Gugutzer sees self-tracking as the objectivation of the zeitgeist, as selftracking "constricts the abstract phenomenon of Zeitgeist to a concrete, meaningful impression, through which Zeitgeist becomes graspable and manageable, which also means analyzable" (Gugutzer 2016, p. 178). Analytically, from a technical perspective, the conflation of virtual and material-physical worlds appears as a significant feature of self-measurement. From a semiotic perspective, the practice of self-measurement refers to a cultural understanding that points to moments of self-optimization in the sense of cybernetic capitalism in the control society. Duttweiler and Passoth (2016) point to the interlocking of self-optimization and control when they state, "It is foreseeable that the modes of self-measurement that exert more or less 'gentle' coercion on individuals to measure themselves and to forward their data to these institutions will become increasingly central" (Duttweiler and Passoth 2016, p. 19). Quantitative "knowledge about oneself becomes the vehicle of the change of a type of rule and government: from foreign to self-government" (Vormbusch 2016, p. 55). In terms of zeitgeist as a framework for analysis, self-measurement "does not deal with the individual but with *collective* problematic situations that manifest themselves individually" (Vormbusch 2016, p. 51, e.i.o.). From this perspective, "self-measurement practices [...] are related to specific cultural problems of contemporary capitalism: the cultural underdetermination of lifestyle practices and the economic uncertainty about the value of the self, especially as an object of competition and performance" (Vormbusch 2016, p. 51).

Self-measurement makes it possible to generate data about one's own body and consequently to optimize it. Self-measurement can be read as a "valorization practice of previously unquantifiable aspects of the self" (Vormbusch 2016, p. 47). In the course of this subjectification process, the "self is stylized as the central agent" (Strübing et al. 2016, p. 272). The individual turns himself and his performance into an object to be optimized. In this process of the digital measurement, data possess an interpellative power. Data have the "inherent 'aura' of the objective" (Schaupp 2016, p. 68). Through the installation of self-tracking as a specific form of interpellation, a "conversion from feeling to measuring" (occurs Strübing et al. 2016, p. 276). This conversion from feeling to measuring entails "a serious conversion in the self-relation" (ibid.), which "experiences a decisive dynamization with the new digital media" (ibid.). Strübing et al. (2016) state about their research results:

In general, it is noticeable that the self-measurers we studied orient themselves – with regard to both the process and the goals of self-measurement – to standards and norms that are discursively conveyed and predominantly represented in the apps and devices: A BMI between 20 and 25 is considered 'normal,' three liters of water per day appear to be healthy, the pulse while running should be at most 220 beats minus the age, 10,000 steps per day are considered moderate exercise, food is classified by apps as healthy (green), in moderation (yellow), unhealthy (red), etc. (Strübing et al. 2016, p. 280).

The "reference to one's own body becomes even more indirect" (ibid.), as the activity trackers unfold an interpellative power. This interpellative power inscribes itself in the body when the activity trackers represent *"bodily communication partners* for the actors" (Gugutzer 2016, p. 165, e.i.o.). An *"authority of things"* is established that possesses "a *bodily gripping power"* (Gugutzer 2016, p. 167, e.i.o.): "The power is bodily gripping insofar as the shapes, colors, etc. of the things produce a perceptible resonance in the self-trackers" (ibid.). Thus, the "hum of the wristband [...] can trigger a fright or awaken a guilty conscience because it means having moved too little" (ibid.). Reichert speaks here of an "extension of medial technologies into the subject" (Reichert 2016, p. 189).

Within the framework of digitally supported self-measurement, a unidirectional orientation of communication occurs. With reference to Baudrillard, it could be said that this unidirectional, non-dialogical communication leads to a stabilization of power relations. The individual does not have the possibility of dialogical interaction and participation but reacts to the interpellative signals of the activity trackers: "The rules of the game are programmed unilaterally and can only be executed by the users, but can no longer be modified themselves to influence the structure and course of the game" (Reichert 2016, p. 188). Like social networking sites, activity trackers span a normative space of possibility. Accordingly, Strübing et al. (2016) point out that "the practice of self-measurement fits seamlessly [...] into the growing number of net-based services, especially in the field of social media" (Strübing et al. 2016, p. 282), "in which users make public data about themselves for the sake of limited benefits, with which the operator-entrepreneurs realize high profits in the advertising business" (ibid.).

While digital media in Cyberspace function as liberation technologies, digital media in the self-tracking of the SNS universe evoke control effects: "The borderline objects of digital self-measurement on online platforms are located at the interface of reflexive self-thematization and social feedback technologies" (Reichert 2016, p. 191). The fit of self-tracking into the logic of the SNS universe is evident when, for example, training achievements can be posted to online communities such as Facebook via one click. The fixation of the individual in the SNS universe is thereby performed as a collective experience. In this way, via

> *Live tracking* [...] athletes share their position, determined through geo-tagging, with friends in social networks (Facebook, G+, Twitter) and on various fitness portals. By means of a cheering function available on Runtastic, users can be motivated in real-time connections with cheers of up to five seconds in length (Reichert 2016, p. 193).

Self-measurement promises possibilities for self-optimization and thus a symbolic increase in the competitiveness of the entrepreneurial self. From this perspective, self-tracking appears as a way out to counter the precariousness resulting from the competition of entrepreneurial selves. As Vormbusch (2016) points out,

> that self-measurement as a life management practice promises to make the experiences of uncertainty in the fields of economic competition (performance), bodily and mental health, and emotional management workable using quantifying self-observation (Vormbusch 2016, p. 47).

Against the background of the interpellative implications and the promise of processing of precarity, it can be assumed "that the media of self-measurement provide a technically mediated framework that does not merely subjugate or oppress users, but transforms them in a certain way and makes them productive" (Reichert 2016, p. 185). It is possible to speak of a subjectification of the individual according to the normative claims and interpellations of the neoliberal zeitgeist in the practice of digitally supported self-measurement. The strengthening of one's performance via self-optimization through self-tracking promises a strengthening of the individual in the competition of entrepreneurial selves. Behind the promise of security through self-optimization, a circular logic reveals itself. The self-enhancement is based on an inherent competitive logic that comes to light, among other things, when, for example, training results are ranked. This can happen intra-individually, for example, when one's training performance is compared and ranked by a self-tracking app. Inter-individual competition is also possible when training results are related to the training results of other individuals. With reference to Boltanski and Chiapello, Vormbusch attributes the discursive identity of the practice of self-tracking to the "connection between artist critique and new management" (Vormbusch 2016, p. 48). Thus, self-tracking is not conceivable without "the collective claims to successful identity" (Vormbusch 2016, p. 48), "which have been developed in self-help, protest, and emancipation movements in recent decades" (ibid.).

Via the narrative of the entrepreneurial self, a neoliberal concept of freedom is established. The order of a control society organizes itself via cybernetic feedback processes. Following Crouch, Stalder (2016) sees features of post-democratic forms of society: "In the spirit of cybernetics and compatible with the structures of post-democracy, people are to be moved in the direction determined by experts via changes in the environment, while at the same time being given the impression of acting autonomously and freely" (Stalder 2016, p. 229). The appearance of freedom and self-determination arises via the metanarrative of the market in which the individual is free to act. This localization of the individual in the normative possibility space of neoliberalism effects a fixation of the individual. Accordingly, Wild (2017) notes that in "times of the Internet [...] a social form that is not only expressed in the medial seems to be emerging" (Wild 2017, p. 82), "in which both practices and the spectacle coincide" (ibid.). The spectacle, one could continue, represents the discursive staging of the entrepreneurial self. Through this, "surveillance gains legitimacy because it becomes part of the spectacle and enter-, info-, edutainment" (ibid.). Through "the media seductions of commercialized spectacles or new digital possibilities" (ibid.), the "possibilities of surveillance [...] (Big Data, Learning Analytics, Database Marketing)" (ibid.) fall into the background.

*The ambivalence between subjectifying neoliberal freedom, which relies on the flexibility and activity of the individual, and the individual's fixations by the power-strategic dynamics of the control society effectuates a neoliberal culture of the digital within the SNS universe. This culture of the digital extends to the material-physical world in the sense of post-digitality. The cultures of the digital exhibit a range between postmodern narratives of freedom and neoliberal self-optimization. These digital cultures exist alongside and against each other and are fighting a battle over the symbolic order of the Internet.*

The lines of conflict inscribe themselves on the Internet from the material-physical world. They are a legacy of the protest movements of the 1960s and 1970s and the alternative culture of the 1980s and the establishment of the neoliberal metanarrative since the 1980s. Actors are embedded in these cultural interpretations offered by the Internet. *In the sense of the difference-theoretical approach of hyper-culture, the individual becomes the vanishing point of cultural practices in the digital age. The challenge is to gain a reflexive approach to the Internet's complexity as a space of cultural practices.*

From the perspective of media education, the challenge arises to enable a sovereign, reflective approach to the cultural practices evoked by the Internet. This media pedagogical challenge can be analytically processed with the concept of media Bildung. Thus, a change of levels is carried out within the framework of this thesis. While the first three chapters are based primarily on cultural theory and discourse analysis methods, which draw integratively on media and epistemological analyses, the following chapter mostly adopts a media education perspective. In this way, the cornerstones of postmodern media Bildung can be outlined, which makes it possible to adopt a critical-reflexive relationship to the digital cultures, which were discussed in the first three chapters.

# The University in the Digital Age as a Place of Postmodern Media Bildung

**4**

## 4.1 The Unconditional University as a Space of Postmodern Reflection

Universities have been central to the emergence of the Internet and the process of digitization. Castells (2005) points out that the military relied on the involvement of scientific freedom in the development of the ARPA-Net, and scientists were given the opportunities to "turn their ideas into manageable ideas in research and scholarship" (Castells 2005, p. 31). Castells stresses that "all the crucial technological developments that led to the Internet emerged around government institutions, large universities, and research centers" (Castells 2005, p. 32). At the same time, the university represents a historically evolved site for critically questioning social practices and transformation processes such as digitalization. Fisch (2015) points out that universities historically have a special significance in Western society's symbolic order., Since "their medieval period of origin" (Fisch 2015, p. 7), universities had "signs of power such as scepters and signs of confirmation such as seals, which a university and its officials were entitled to wield" (ibid.). These symbols of power marked the special rights that distinguished the university. "The core of the universities' special right lay in the fact that it dealt with several fields of knowledge at a high level according to self-imposed rules" (Fisch 2015, p. 7). Stalder (2016) states that "sciences [...] were the first major social field [to] acquire comprehensive cultural autonomy, that is, the ability to determine for themselves the meaning that was binding for them" (Stalder 2016, p. 152). Steinert (2007) problematizes that the university "for the vast majority of its history has not been the site of liberated and liberating thought, but rather the haven of orthodoxy, often enough of reaction" (Steinert 2007, p. 18). At the same time, Steinert acknowledges that there is also the "founding myth of Bologna" (ibid.), "according to

© Springer Fachmedien Wiesbaden GmbH, part of Springer Nature 2023
D. Kergel, *Digital Cultures*, https://doi.org/10.1007/978-3-658-35250-9_4

which this first university in Europe was an association of students who sought out their teachers" (ibid.). Steinert states an ambivalence of university. This ambivalence results from the social positioning of the university:

On the one hand, university is discussed as a place where knowledge *of* and *about* society is critically examined and questioned.
On the other hand, a university is a place where future specialists and managers are trained. Therefore, the university must meet the requirements of a labor market for academics and ensure the future employability of its students.

This ambivalence of the university also shapes the discussion of the university's social meaning and function in the digital age. To take an analytical look at this ambivalence, the university's epistemological dimension will be discussed below. In the course of this, possibilities of a postmodern media Bildung will be explored, which unfold in the university educational space.

With reference to Derrida, university can be understood as an educational space that makes it possible to critically engage with digitalization or with social transformation processes in digitalization. The starting point is Derrida's concept of the unconditional university, which he developed in a lecture given at Stanford University in 1998. Derrida states that the confrontation with the knowledge or the possibilities of secure knowledge is a constitutive characteristic of the university: "The university makes truth *its profession* – and it *professes the* truth, it takes a vow of truth. It publicly declares and vows to live up to its unqualified commitment to truth" (Derrida 2015, p. 10, e.i.o.). Accordingly, the university is defined as a space characterized by the fact that everything can be questioned. This also applies epistemologically to the act of questioning itself. According to this conception, the "university [...] would thus also have to be the place where nothing is out of the question" (Derrida 2015, p. 14). Through performative questioning, a significant feature of postmodern epistemological is realized. This postmodern approach to knowledge allows the university to become a strategic epistemological critique: By questioning everything, the university undermines truth claims. The university becomes a space of reflexive postmodern resistance. Derrida calls this reflective resistance 'unconditional' resistance. According to Derrida, this unconditional questioning of everything, this unconditional resistance functions as a university's constitutive principle. The "unconditional resistance [could] put the university in opposition to a whole range of powers" (Derrida 2015, p. 14). Derrida sees universities in opposition to "media, ideological, religious, and cultural powers, etc., in short, to all the powers that constrain the democracy to come and remain to come" (Derrida 2015, p. 14). University acquires an ethical basis through its function as an instance of democracy. The university keeps the democratic discourse space open for dialogical processes of negotiation through critique of knowledge. Derrida points out that the unconditional university does not exist as a place of resistance but considers it relevant as a symbolic place of resistance:

> As we know only too well, this unconditional university does not *de facto exist*. Nevertheless, in principle and according to its admitted vocation, its declared essence, it should be a place of ultimate critical – and more than critical – resistance to all dogmatic and unjustified attempts to seize it (Derrida 2015, p. 12).

University is the space of scientific critique. Bourdieu (2015) sees to this form of the scientific critique of knowledge – and here he is in an argumentative line with Derrida – the potential of a 'counter-power,' without which there is "no real democracy" (see Bourdieu 2015, p. 158). Bourdieu sees in the systematic doubt of scientific epistemological critique a paradigmatic model of discourse: "I believe that all the world would gain much if the logic of intellectual life that of argumentation and refutation were to extend to public life" (Bourdieu 2015, p. 159). This form of the scientific critique of knowledge in the context of social and political self-understanding discourses is systematized by Steinert (2007). Steinert uses the term 'intellectual politics.' In the context of intellectual politics, it is "[t]he tasks of intellectuals [...] to be responsible for the reason as a whole and thus to represent *general* interests for which no single-interest lobby can be found" (Steinert 2007, p. 25, e.i.o.). According to Steinert, intellectual politics is defined, by its advocacy of civil and human rights and by its questioning of "domination and its form" (ibid.). Such a perspective on the university as a space for the formulation of intellectual politics represents "not least a reminiscence of the sixties and the student protest movement" (Hirsch 2007, p. 241). Philosophical politics can build on the prominent position of science, which it is accorded in the context of social processes of self-understanding. Thus, 'scientifically secured knowledge' is of central importance, especially in modern societies: "Modern societies are characterized, among other things, by the fact that in them a special kind of knowledge, precisely scientific knowledge, is regarded as the final authority in the solution of problems of all kinds" (Weingart 2003, p. 15). In a society that, like bourgeois society, sees itself as secular, rational (see Krämer-Badoni 1978) and knowledge-based, science is assigned the task of producing valid knowledge about the world. Such knowledge can provide orientation patterns for political decision-making. Steinert's approach to intellectual politics also corresponds to this approach, whereby Steinert provides a power-critical, emancipative interpretation of science. Steinert's reflections show the university's potential as a postmodern space of subversion that offers a space for questioning conventionalized self-evident truths: "Incidentally, intellectual politics was and is always firstly anti-authoritarian politics and secondly the politics of the utopian. Knowledge-promoting politics must let the self-evident become self-evident [...] It will therefore reverse acts of domination" (Steinert 2007, p. 26).

From the perspective of scientific theory, scientific knowledge can be defined because it represents systematic, intersubjectively ordered knowledge. Scientific knowledge it is logically argued and thus intersubjectively comprehensible. Logical argumentation orders and systematizes knowledge. Data and facts are put into a meaningful relation to each other. Appropriate methods generate these data and facts. These methods are theoretically or methodologically sound. In other words, scientific knowledge is methodologically sound knowledge. Scientific methods and scientifically generated knowledge exhibit quality characteristics of science, such as reliability, validity, and objectivity. Perhaps the most central feature of scientific knowledge, however, is its epistemic stance. Science adopts a doubtful perspective towards conventionalized knowledge by critically questioning its validity. Scientifically secured knowledge does not represent a final 'true knowledge.' Instead, the state of scientific knowledge appears to be precarious: "The game of science has in principle no end: whoever decides one day not to examine the scientific propositions further, but to regard them, for instance, as finally verified, steps out of the game" (Popper 1973, p. 26). Scientific knowledge is always precarious and never final. Scientific progress manifests itself in the realization that one has succumbed to an error: "Scientific progress does not lie in the fact that more and more new experiences come together over time; nor in the fact that we learn to use our senses better" (Popper 1973, p. 224). Science takes place in research that means, in a methodologically founded critique of knowledge. Such research is constitutively infinite:

> Admittedly, we do *not know, but we guess*. And our guessing is guided by the unscientific, metaphysical (but biologically explicable) belief that there are regularities which we can unveil, discover [...] But these often fantastically bold anticipations of science are clearly and soberly checked by methodical verification. Once established, no anticipation is dogmatically held on to; research does not seek to defend it, it does not want to be right: with all the means of its logical, mathematical, and technical-experimental apparatus, it tries to disprove it (Popper 1973, p. 223, e.i.o.).

Instead of a teleological understanding of science, science can be defined as a curious and, epistemic-critical relationship to the world. Scientifically assured knowledge is constitutively characterized by recognizing the possibility that phenomena are structured differently than assumed. From this perspective, scientifically secured knowledge is always uncertain knowledge and an expression of a postmodern attitude towards knowledge. Knowledge is strategically questioned. The questioning, epistemic-critical attitude of science is a significant feature of a postmodern perspectivization on the world: Lyotard sees science as being permeated by a (postmodern) doubt:

> Science does not expand by means of the positivism of efficiency. The opposite is true: working on a proof means searching for and "inventing" counterexamples, in other words, the unintelligible; supporting an argument means looking for a "paradox" and legitimating it with new rules in the games of reasoning. In neither case is efficiency sought for its own sake; it comes, sometimes tardily, as an extra, when the grant givers finally decide to take an interest in the case. But what never fails to come and come again, with every new theory, new hypothesis, new statement, or new observation, is the question of legitimacy. For it is not the philosophy that asks this question of science, but science that asks itself (Lyotard 1983, p. 54).

Against the background of the performative and postmodern doubt of science and regarding Derrida's concept of the unconditional university, the university can be understood as an educational space that methodologically grounds the postmodern critique of knowledge: Instead of the establishment of hegemonic knowledge or metanarratives, the producing of scientific knowledge takes place as a constant doubting. Truth functions as an ideal to which it is necessary to approach, but which cannot be attained, since knowledge is always undermined by the doubt: "Although truth and probability are unattainable for it [science], the intellectual striving, the drive for truth, is probably the strongest impulse of research" (Popper 1973, p. 223).

The scientific process of systematic doubt is to be understood as a collaborative process. Research is not an act of a singular individual who generates knowledge in a lonely study room. Instead, research – understood as the practical dimension of science – is understood as a collaborative epistemological process. The producing of scientific knowledge can be understood as common epistemic project, in which truth postulates are systematically challenged with the intention of falsification. One consequence of this argumentation is that there is no single fixable originator of knowledge. This is an insight that underlies the open-source movement. Knowledge is generated and critically questioned in a dialogical intersubjective exchange which is based on a social constructivist performative process. In the course of collaborative research processes, individual knowledge becomes fluid. Accordingly, Giddens (1976) sees merit in Popper's formulation of the principle of falsification, which also strengthened the scientific community:

> Popper's philosophy of science broke fundamentally not only with logical positivism but also with traditional conceptions of science that tend to treat the scientific method

from the point of view of the individual scientist; he conceives of science instead as a collective enterprise, institutionalization of the critical reason (Giddens 1976, p. 165).

*This collective research finds its medial equivalent in the decentralized struc-tures of the Internet.* Against the background of the collaborative dimension of re-search, it is not surprising that the Internet has always been an enterprise of col-laboratively working scientists: the Internet's basic technical structure was created "in the 1960s/70s" (Hartmann 2006, p. 167). This happened in several steps, "which makes it almost impossible to speak of inventors and their concrete achieve-ments here anymore" (ibid.). The Internet was not invented by a scientist but in the process of exploratory exploration that developed successively in the form of col-laborative work. Berner-Lee et al. (1994) point out that the development of the World Wide Web was based on the intention to promote processes of decentralized, collaborative knowledge construction:

> The World Wide Web (W3) was developed to be a pool of human knowledge, which would allow collaborators in remote sites to share their ideas and all aspects of a com-mon project. Physicists and engineers at CERN, the European Particle Physics Laboratory in Geneva, Switzerland collaborate with many other institutes to build the software and hardware for high-energy research. The idea of the Web was prompted by the positive experience of a small ‚home-brew' personal hypertext system used for keeping track of personal information on a distributed project. The Web was designed so that it was used independently for two projects, and later relationships were found between the projects, then no major or centralized changes would have to be made, but the information could smoothly reshape to represent the new state of knowledge. This property of scaling has allowed the Web to expand rapidly from its origins at CERN across the Internet irrespective of boundaries of nations or disci-plines (Berner-Lee et al. 1994, S. 792).

From a discourse-analytical perspective, an updating of the semantics of freedom in cyberspace can be observed. The Web makes it possible to transcend boundaries and thus to work collaboratively. Berner-Lee et al. see an "exciting future" (Berner-Lee et al. 1994, p. 797), as an "incredible diversity of information" becomes acces-sible through the connecting lines of the World Wide Web. The ephemeral structure of the digital corresponds to the ephemeral dynamics of scientific knowledge pro-duction. Collaborative science, one might conclude against the background of the Internet's roots, unfolds its potential through the decentralized, polyphonic, and poly-directional possibilities of the Internet. Accordingly, the decentralized, poly-directional, and polyphonic structure of Web 2.0 media also enables collaborative research. For example, Berry (2014) holds that "computer code enables new com-munication processes" (Berry 2014, p. 53): "[W]ith the increasing social dimension of networked media comes the possibility of new and exciting forms of collaborative thinking" (ibid.).

In this context, Berry refers to the role of the research community. This research community can be organized through digital media, thus enabling a spatially as well as temporally decentered structure: "The situation is reminiscent of the medieval notion of a *universitatis*, but digitally transformed, as a society or association of agents who, thanks to the mediation of technology, can think critically together" (Berry 2014, p. 54, ed.). Nitsch (2007) sees media change as having the potential to support the critical and subversive nature of scholarship. Thus, "new communication technologies allow for cheaper, faster, and more difficult to control critical science, and in addition to that informed political communication between spatially and professionally dispersed scholars and students" (Nitsch 2007, p. 204). A digitally organized research community substitutes a romantic image of the scientist as a researching hero who embarks on the adventure of knowing and not knowing all by himself. The construction of scientific knowledge can be understood as an effect of a rhizomorphic research community. Participatory, online-based media such as blogs, wikis, and podcasts have the potential of collective authorship, replacing the book as the leading medium of scholarship. Schwalbe (2011) points out that "[n]o longer the book is the central medium of communication" (Schwalbe 2011, p. 139) in dealing with knowledge, "but a participatory, interactive, globally networked medium [...] is establishing itself as a ubiquitous medium of communication, dissemination" (ibid.). The media change changes the discursive (self-)understanding of science: science blogs mix scientific discussions, science journalism, and scientists' self-promotion strategies. With the logo "Write research together," collaborative writing tools like Authorea push the concept of collective authorship. In other words: in the digital age, science makes constitutive use of digital media:

> Every research project goes through generating ideas, gathering information, or publishing results, and every teaching project involves processes of imparting expertise or reflection. Digital media are used en masse to manage the processes, and digital objects are generated en masse. Communication processes and access to materials also proceed digitally (Seiler-Schiedt 2013, p. 269).

In the scientific field the media change redefines scientific practices, and places changed demands on academics' media competence. The methods of knowledge construction are changing. Retrospectively, such a change in scientific practices can be reconstructed with the establishment of the Gutenberg Galaxy. Stalder (2016) points out that new conventions emerged in the scholarly field as an effect of the rise of book production: Books were marked with the data of authors, printers, publishers, and the year of publication. Data and methods emerged that enabled the referencing of text passages and also define scholarly citation to this day. A text-oriented scholarly culture was made possible: "scholars could legitimize the claim to be searching for new knowledge by pointing out errors or gaps in existing texts" (Stalder 2016, p. 103). In this way, "[i]n the developing scientific culture [...] the close interlocking between existing and new material is not only positively valued, but mandatory as a method of argumentation" (ibid.). With the digitalization of society, the media dimension of science is also changing. This means that traditional scientific techniques such as citation are also at stake. The question arises, how citation techniques established in the context of the Gutenberg Galaxy can be extended to the new scientific media and digital artifacts:

How can one appropriately cite a discussion that originated in a blog, YouTube video, or chat room?
Is Wikipedia citation-worthy or not?

These questions mark a media break in the field of science. Against the background of the scientific quality criteria of objectivity, reliability, and validity, it is necessary to negotiate new scientific techniques, which necessitates a redefinition of the demands placed on academic media competence. Against the background of these considerations and regarding Derrida's model of the unconditional university, the university can be conceptualized as space,

In which new digitally supported forms of knowledge construction are developed and.
a critical reflection and analysis of the process of digitalization and the accompanying social transformations, must be carried out.

This is how digitalization is changing

> also significantly shape expectations for twenty-first-century science, and there are increasing calls to insist that the historically, culturally, and socially influential aspects of digital data practices be systematically addressed – linked to the goal of anchoring them in future scientific cultures and epistemologies of data generation and analysis (Reichert 2014, p. 11).

Against the background of these considerations, the university becomes a place of critical questioning of digitalization. This questioning is organized via digital media as a collaborative process of knowledge. In this questioning, it is important not to understand digitization as a force of nature that breaks over civilization unchecked and uncontrolled. Society is not at the mercy of digitalization. Rather, it is important to evaluate digitalization as a social, cultural product and discursive construct to relate critically. The interpretation of digitalization as a social product can be critically questioned through scientific strategies of knowledge. Digital media are not only an object of reflection but also a medium. This critical-reflexive and at the same time action-oriented approach to digital media and the challenges of digitalization can be conceptually analyzed using the term media Bildung.

## 4.2 Postmodern Media Bildung

The examination of media Bildung makes it possible to conceptualize a critical-reflexive, action-oriented and self-determined handling of media. In the term media Bildung, two words meet in the form of a determinative compound, which are of central importance in a knowledge-based society in times of increased digital change. At the same time, media and education represent two terms whose conceptual-analytical framing does not seem feasible against the background of diverse epistemological approaches. This indeterminacy of the central terms due to their conceptual breadth makes it necessary to work out, which understanding of media and education underlies them.

The digital transformation is changing how individuals form a relationship to the self and the world through their use of media or by acting in augmented reality. The first conceptual setting is to understand the term education as constructing a self/world relationship. This position draws on Humboldt's understanding of education as Bildung, which emerged in the context of German Idealism's subject philosophy (see Musolff 1989; Kergel 2017a). Conceptually, the term subject develops analytical access to how the individual develops a self/world relationship. This self/world relation is thereby to be understood as an emotional-cognitive self/world relation. In the following, this thesis will be developed that the model of becoming a subject is to be understood as an epistemological foundation of Bildung. Consequently, as Bildung, education is to be understood as processual, performative, and infinite: As a form of becoming a subject, education is part of the never-completed process but is constantly becoming. From a praxeological perspective, this becoming is continually actualized by the subject's actions.

Before education can be placed in relation to media Bildung and digitization processes and their educational implications, an epistemological classification of education appears to be relevant. Only against the background of an epistemologically founded understanding of education can a scientifically accurate account of media Bildung be formulated. In the following, the epistemological foundation of education will be based on an 'integrative understanding of Bildung (see in more detail Kergel 2017a).

## 4.2.1   From the Ego to Bildung

The subject is defined by the fact that the individual relates to him/herself, perceives him/herself. In addition to that, the subject can reflect on him/herself and his/her integration into the social context, thus constituting a self/world relationship: "Already in antiquity, the subject is formed by relating to him/herself" (Wulf 2007, p. 36). The subject is not constituted exclusively through a reflexive self/world relation but includes an emotional dimension. In German Idealism, the subject formulates itself as an epistemological instance, using the figure of the ego. In the "Critique of Pure Reason," first published in 1781, which can be read as the founding document of German Idealism, Kant states: "The: I think, must be able to accompany all my ideas" (Kant 1956, B132). Fichte increasingly elaborates on the autoreferentiality of the subject. He thus prefigures constructivist perspectives on learning processes at the epistemological level:

> I was the one who was **doing, and the product of** doing, the thing that was doing and the thing that was produced by doing, and the thing that was doing and the thing that was doing were the same. – is at the same time the acting, and the product of the action; the active, and that which is produced by the activity; action and deed are the same; and therefore the: **I am, is the** expression of an act; but also the only possible one (Fichte 1971, p. 96, e.i.o.).

Following Kant and Fichte, Hegel conceptualizes a processual model of the subject, conceiving the subject as epistemology in motion. In the "Phenomenology of Spirit" of 1807, Hegel's 'first major work,' Hegel sketches a genetic version of the subject. As a reflexive-critical instance, the ego is constituted in the course of the engagement with the world. Starting from 'thoughtless consciousness' (Hegel 1952, p. 185), the individual forms in engagement with the world an 'observing consciousness' (Hegel 1952, p. 189, see also 351ff.). A reflexive relationship to the self and the world is worked out in thinking operations/abstractions such as the construction of comparisons or "analogies" (Hegel 1952, p. 190) and the formulation of hypotheses about the nature of the world (see ibid.). The gradual formation of cognitive abilities in the course of engagement with the world described by Hegel can be read as a prefiguration of socio-constructivist approaches and models of cognitive development, which unfold a.o. in Piaget's model of genetic development theory (see Kergel 2011a, b).

Humboldt stands in the context of these subject-philosophical considerations. Thus, Musolff (1989) points out that "following Kant's epistemology, ethics, and critique of judgment, Humboldt [...] both researches the fundamental aspects and constituents of the human condition and seeks to uncover its political-social implications" (Musolff 1989, p. 119). The philosophical anthropology of German Idealism, from which Bildung-theoretical premises successively develop – "The concept of Bildung does not belong to the peripheral, but the fundamental concepts of German Idealism" (Zantwijk 2010, p. 72) – is updated and further differentiated by Humboldt.

Humboldt's initial thesis is that the self/world relationship is not given statically but develops, expands, and is reinforced in genetic dynamics. The description of these constitutional dynamics of self/world relations represents Humboldt's genuine contribution to German Idealism's epistemological concept of the subject.

> The human being only wants to strengthen and increase his nature's powers to give his being value and duration. But since mere force needs an object in which to exercise itself, and mere form, pure thought, needs a substance in which it can persist while expressing itself, a man also needs a world outside himself. He is not interested in what he acquires from it, or in what he produces by virtue of it, but only in his inner improvement and ennoblement, or at least in the satisfaction of the inner restlessness that consumes him. Viewed purely and in its final intention, his thinking is always only an attempt of his spirit to become comprehensible before itself, his acting an attempt of his will to become free and independent in itself (Humboldt 1980, p. 253).

Humboldt describes the modalities through which a self/world relationship develops. According to Humboldt, the human being develops through engagement with the world. In the course of this engagement, the individual as subject expands his knowledge or learns more about the world. The subject is part of the world. By learning about the world of which the subject is part, the subject also learns about himself and his 'being-in-the-world.' We can speak of a successive unfolding of a self/world relationship, which can be conceptualized by the term Bildung. Humboldt takes the term Bildung from the "contemporary natural philosophical and specifically biological discussion about the formation of beings in nature, especially the formation of living beings" (Zöller 2010, p. 179 f.). He applies the term Bildung to the genetic unfolding of human beings. Analogous to the inherent lawfulness in which nature develops, Humboldt assumes that humans also develop or 'form' themselves in the course of learning-based Bildung processes. Bildung and therewith takes place 'quasi naturally' in the sense of an anthropological constant.

To work through the dynamics of movement that enable recovery, Humboldt draws on the concepts of power and freedom. Humboldt takes the concept of power from scientific discourses and, here in particular, from the physics of his time. "Power is the (ontological) object of Bildung" (Zöller 2010, p. 181). The Bildung process is oriented towards "a dynamic conception of nature as a product of physical forces" (ibid.). The power of man unfolds in his confrontation with the world. In this confrontation, power manifests itself and, thus, metonymically, the nature of man. Humboldt posits an ontologization of the power that drives the Bildung process. Power requires spaces of unfolding. Force is dependent on freedom or spaces of freedom in order to unfold appropriately. This necessity of freedom for Bildung constitutes epistemologically the normative dimension of Bildung and consequently of education. Thus, according to Humboldt 2010, for "Bildung [...] freedom is the first and indispensable condition" (Humboldt 2010, p. 22). Through the concept of freedom, Bildung becomes an ethical process: "Freedom, therefore, means the concrete process by which man chooses a fundamental intentional act the way of life appropriate to the realization of his individuality" (Musolff 1989, p. 125). The interplay of power and freedom forms the normative dimension of Bildung: "Based on the natural-philosophical trinity of Bildung, power and individuality transposed into cultural philosophy, Humboldt develops a normative conception of human development, at the center of which is freedom" (Zöller 2010, p. 184).

From this perspective, Bildung is also an ethical enterprise, as it normatively demands a specific, value-oriented relationship of the individual or subject to the world. Rohlfs et al. (2014) point out that Bildung should be understood as an approach developed "with the aim of liberation" (Rohlfs et al. 2014, p. 12). Thus, "Bildung [...] cannot be reduced [...] to the development of professional competencies" (ibid.) but has an emancipative moment.

Due to the enthronement of the subject, which – driven by power – requires experiences of freedom to educate itself, the process of Bildung is to be distinguished from that of subjectivation. While social values and norms, hierarchies, and dependency relations are inscribed in the individual in the course of subjectification dynamics, Bildung is characterized by a critical-questioning relationship to such value-settings (see Krückel 2017, p. 60 f.). Ethical implications are also embedded in this understanding of Bildung. The ethical implications of the anthropological dimension of Humboldt's concept of Bildung become clear from the five characteristics of the "classical concept of Bildung" listed by Hastedt (2012) (this classical concept of Bildung was significantly developed by Humboldt in addition to Herder, see Hastedt 2012, p. 9): "self-Bildung, shaping and development of the whole person, anthropological neediness and 'growth,' enhancement of individuality with simultaneous supra-individual commitment and overcoming alienation" (Hastedt 2012, p. 9).

Against the background of difference-theoretical concepts of the subject, which understand the self/world relationship as the result of social inscriptions, understanding the subject formulated here stands out: the consideration of the modes of social inscriptions is taken into account. The consideration of the modes thereby analytically highlights the ethical level of Bildung: Does the subject form itself in the confrontation with the world or social dynamics, accesses, etc., or is it subjectivized?

A hermeneutic-understanding approach assumes an aesthetic dimension of education, which, following Dilthey (see 1962), focuses on *experience* as (prereflexive) education/learning. From this perspective, as an agent of cognition, the 'I' is not the manifestation of a purely reflexive cogito, as might be claimed with reference to Descartes. Instead, as a manifestation of a self/world relation in a social context, the subject is always situated in the dynamics of the subject's social *experience*. This aesthetic dimension of Bildung finds its epistemological precursors in the reflections of Novalis and Schlegel on the experience of infinity, among others. Wittgenstein summarizes it as the end of logic: "Whereof one cannot speak thereof one must remain silent" (Wittgenstein 1963, p. 115). The subject of the Bildung process *experiences* learning and education and constitutes itself as a subject in this *experience*.

## 4.2.2   Empirical Opening of the Educational Theory Model

Regarding an empirical validation of these theoretical Bildung considerations, the question arises as how concepts such as power and freedom can be appropriately operationalized for empirical research: Exploratory curiosity, which can be understood as an expression of power in Humboldt's sense, forms a starting point for the empirical opening of the model of an Bildung theoretical approach towards media. Without a form of exploration, the individual's cognitive and emotional development seems inconceivable, which suggests that explorative curiosity, like power, should be understood as an anthropological constant. Exploration, or exploratory curiosity, is manifested in curious, exploratory behavior toward the world (see Gibson 1998). From this perspective, explorative curiosity can be defined as an epistemological curiosity that drives Bildung processes.

Explorative curiosity requires spaces of possibility or freedom to unfold so that the subject can 'strengthen and increase the forces of its nature.' These spaces of possibility release potentials for (postmodern) learning.

Concerning the construction of empirical indicators that refer to the existence of Bildung, Bandura's concept of self-efficacy (1977) can be adapted: Explorative curiosity requires that the individual learns and, in the course of this learning, experiences self-efficacy and thus develops experiences of self-efficacy. Self-efficacy can be defined as the learner's conviction that appropriate behavioral outcomes can be realized in a situation. Self-efficacy expectancy refers to the subject's expectation that he or she possesses the competencies to cope with a challenging situation. Accordingly, successful actions reinforce self-efficacy expectations with these self-efficacy experiences based on exploratory curiosity, the individual experiences a Bildung process.

Explorative curiosity and self-efficacy expectations can be defined as the modes or experience dimensions of Bildung. In the course of empirical research, it must be examined whether indicators can be found in the engagement with the world or in the construction of self/world relationships that point to the presence of explorative curiosity and self-efficacy expectations. From this perspective, Bildung can be defined as a positively connoted self/world relationship based on explorative curiosity and self-efficacy experience, which is produced performatively.[1] Bildung unfolds constitutively in learning processes.

### 4.2.3  Bildung Learning

In the course of learning processes, the subject constructs knowledge about him/herself and the world and develops his/her relationship to the self/world. From this perspective, knowledge and Bildung have a specific relationship to each other:

> Even though Bildung is based on knowledge and always includes the transmission of knowledge, it is neither about knowledge as knowledge nor about the mere adoption of knowledge that is recognized as objective; rather, Bildung aims at a specific relationship to knowledge that can be described as a particular relationship to the self and the world (Ricken 2006, p. 321, footnote 200).

Bildung, knowledge, and thus also *learning* as a process of knowledge construction stand in a specific relationship to each other. Schulze (2007) points out that the "semantic independence and terminological segregation of the concept of Bildung have led to its isolation from the conceptual field and subject area of learning and

---

[1] Positive connotation here means a basic mood with which the individual constructs or forms self/world relationships. A negatively connoted self/world relationship would in turn be characterised by experiences of fear towards the world and a deficit-oriented self-relationship.

has prevented a differentiating elaboration of the learning processes for which it stands" (Schulze 2007, p. 154). The integrative merging of the concepts of learning and Bildung leads to a shift in emphasis. Instead of differentiating Bildung from learning, it is a matter of differentiating the specific forms of learning that also make Bildung possible. The modes or qualities of learning are taken into account. Bildung takes place in a specific form of learning. Against the background of the epistemologically derived understanding of Bildung, learning takes place through explorative curiosity and the emergence of self-efficacy expectations. Such learning actualizes the elements that (co-)constitute Bildung and can accordingly be defined as Bildung learning. Bildung learning can take on different characteristics and unfold in different forms of learning. Against the background of these considerations, a continuum of Bildung learning can be sketched that ranges from forms of discovery learning to inquiry learning. The continuum's axis is characterized by the increasingly reflexive relationship of the learner to his or her learning. In general, these forms of learning are characterized by a curiosity for knowledge, which can be defined from a learning theory perspective as forms of explorative curiosity:

In the course of discovery learning, knowledge is constructed by the learner in active engagement with the environment. Discovery learning can be observed in infants, among other things, through exploratory behavior (see Gibson 1998). It can be interpreted in terms of an anthropological constant.

Self-regulated learning has a mediating function within the learning continuum by leading from intuitive learning strategies to a reflective approach to learning: Self-regulated learning is defined by an increased meta-reflection on learning (see Zimmerman 1990). In the sense of Piaget (1970), this meta-reflection requires the acquisition of formal-operational competencies and is consequently linked to cognitive competencies development.

Inquiry-based learning can be defined as learning *according to* scientific standards and *through the* application of scientific standards in the learning process: Inquiry-based learning is also characterized by the fact that knowledge is constructed theoretically research-methodologically sound manner. For knowledge construction, cognitive strategies and scientific quality criteria are available, which make research-based learning a scientific cognitive process (Fig. 4.1).

All these forms of learning require knowledge to be kept open and would come to an end if the perspective of knowledge were to be closed. Keeping knowledge open in this way is a significant feature of postmodern knowledge: The hitherto generated knowledge or the constructed self/world relation must always be questioned. Without such questioning, explorative curiosity cannot unfold, since instead of

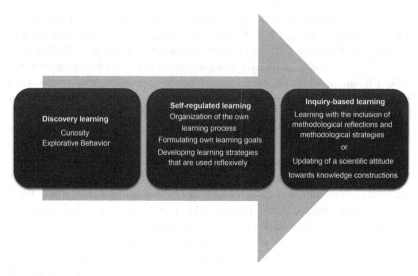

**Fig. 4.1**  Learning continuum of educational learning (own illustration)

learning something new, previously acquired knowledge would only be confirmed. *Explorative curiosity constitutively requires postmodern doubt.* In all these forms of learning, Bildung or Bildung learning or learning takes place. The individual experiences himself as self-effective can unfold his explorative curiosity, and in the process, the 'circle of his knowledge and his effectiveness.'

### 4.2.4  University as an Bildung Space for Inquiry-Based Learning

Inquiry-based learning can be understood as a specific form of Bildung learning. The learning or Bildung space of inquiry-based learning is the university. In higher education, inquiry-based learning takes place as Bildung learning. Humboldt already elaborated the Bildung-theoretical dimension of science:

> It is also a peculiarity of higher scientific institutions that they always treat science as a problem that has not yet been completely solved. Therefore, it always remains in research since the school has to learn only with finished and agreed-upon knowledge. The relationship between teacher and pupil, therefore, becomes quite different from what it was before. The former is not there for the latter. Both are there for science (Humboldt in Hastedt 2012, p. 101).

The university spans an Bildung space as students engage in research-based learning to expand their self/world relationship by engaging with 'as yet unresolved problems' in a scientifically informed way. Such a perspective on university focuses on the unfolding of individual potential against a background of critical reflection through the 'study of science' at university. Fisch (2015), for instance, notes that "[t]he new Prussian-German university, for which the Berlin foundation of 1810 stands paradigmatically [...] set itself the formation and consolidation of the individuality of the single young person as a teaching ideal" (Fisch 2015, p. 53). The focus is on the scientific interest in knowledge, which can be understood as a methodologically sound expression of explorative curiosity (see Kergel 2014). Students should adopt an attitude of inquiry that can "indirectly affect the unfolding of the personality" (Fisch 2015, p. 39). Analogous to Derrida's approach of the unconditional university, according to Humboldt, an unalterable searching characterizes learning in universities: "As soon as one stops searching for science, or imagines that it does not need to be created out of the depth of the spirit, but can be extensively strung together by collecting, then everything is irretrievably and eternally lost to science" (Humboldt in Hastedt 2012, p. 101). This search is tied to the experience of Bildung: "For only the science that comes from within and can be planted within also reshapes the character" (Humboldt in Hastedt 2012, p. 103 ff.). Vinnai (2005) elaborates that the Humboldtian ideal of the 'citizen of the world' unfolds in the university's space of Bildung. With reference to Humboldt, Vinnai understands the university as a "place where individuals and citizens of the world are brought forth or, more precisely, bring themselves forth" (Vinnai 2005, p. 2). The concept of the world citizen addresses the critical-reflexive examination of 'world questions.' It enables a critical view of society: "To become a world citizen means to deal with the great questions of humanity: to strive for peace, justice, the exchange of cultures, different gender relations or a changed relationship to nature" (Vinnai 2005, p. 3).

Bildung takes place in learning processes and is fundamentally open to knowledge. The university provides a space for Bildung: knowledge is questioned in terms of epistemological strategies. Forms of knowledge, world views, and narratives are always precarious, which actualizes the postmodern attitude to knowledge in the field of science. In higher education didactics, the model of inquiry-based learning is currently being discussed (see Kergel and Heidkamp 2016), which is supposed to enable such Bildung learning in the higher education educational space. In this context, media pedagogical aspects are also moving into the discourse on (Bildung based) higher education didactics: In a digital change or a digital age, research-based learning in the university educational space is to be understood as research-based learning with digital media (see Kergel and Heidkamp

2015). Against the background of the Bildung theoretical implications of researching or inquiry-based learning, the question arises of how to conceptualize the relationship between media and Bildung. A condensation of this relationship can be
seen in the concept of a Bildung based media Bildung.

### 4.2.5    From Bildung to Media Bildung

In summary, regarding Derrida, the university can be interpreted discursively as a
'counter-place' or place of resistance. Socially conventionalized narratives, truth
claims, and power structures are critically questioned. Ideally, critical questioning
is part of individual Bildung processes in social contexts that unfold in the university as Bildung space. In this process, critical thinking and individual learning processes interact and are central constituents of Bildung. Against the background of
media change, the media dimension of Bildung is increasingly coming into focus.
The concept of media Bildung takes on a central relevance in the analytical processing of the genesis of self/world relations in the digital age. In general, Bildung
processes are always medially bound. There is no beyond the medial. Against the
background of the media's omnipresence, the question arises as to how media
Bildung can be defined. In the following, we will argue for understanding media
Bildung as a conceptual category that specifically takes an analytical look at the
relation between Bildung and media. Similar usage is also suggested by Marotzki
and Jörissen (2008):

> From the perspective of media Bildung, it is therefore important to analytically rec-
> ognize the reflexive potential of media spaces on the one hand and media forms of
> articulation on the other regarding the aforementioned orientation services and di-
> mensions and assess their Bildung value. The focus here is less on the content of the
> respective media but rather on their structural aspects. The analysis of the medial
> form determinants leads [...] into an analysis of the structural conditions of reflexivity
> processes (Marotzki and Jörissen 2008, p. 103).

Concerning the Bildung characteristics of explorative curiosity and self-efficacy
expectations, it can be researched which factors lead to developing self-efficacy
expectations and allow room for explorative curiosity in dealing with media.

Media Bildung takes place in media action through a (reflexive) action- and
production-oriented engagement with media. In the course of this engagement,
a positively connoted self/world relationship is constituted in the context of the
relation subject/media, in accordance with the normative implication of Bildung.

This relation is not to be understood as dichotomous. Instead, it is in the reflexive and active confrontation that the subject first brings itself forth as such. Accordingly, Spanhel (2011) also proposes to understand media Bildung as a process that manifests itself "in the independent initiation and design of media Bildung spaces" (Spanhel 2011, p. 114).

The abilities and skills that the subject develops in the course of this process constitute media competencies. Media competencies are consequently defined from the reflexive, action- and production-oriented handling of media: Spanhel thus also points out that the term media competence designates the "specific skills [that] we observe in a concrete situation in adolescents or adults in dealing with certain media offerings" (Spanhel 2011, p. 112 f.). The various dimensions of the "applications of those learning tools and patterns of action that have been trained in dealing with media and acquiring media content" (Spanhel 2011, p. 113) can be understood as media literacy.

If this process is researched, media Bildung represents a specific focus within educational research. The approach of media Bildung is defined by the fact that "aspects of mediality in educational science are assigned a systematic, that is, theory-forming and research-guiding value" (Marotzki and Jörissen 2010, p. 100) (see empirically Kergel and Heidkamp 2015; Kergel 2017a).

The media's pedagogical consequences of such an understanding of media Bildung are that didactic strategies enable explorative curiosity and self-efficacy expectation. Such an approach stands in the tradition of constructivist didactics. It must be empirically substantiated through didactic research in the sense of the design-based research approach, scholarship of teaching and learning, or education-oriented evaluation research.

### 4.2.6 Postmodern Media Bildung – Or About the 'Cultural Hacker'

Media Bildung processes are postmodern because – like Bildung processes in general – they require an epistemic opening of knowledge, based on which explorative curiosity and self-efficacy expectations can unfold. To illustrate this epistemological thesis, one can draw on the "postmodern[e] figure of the cultural hacker" proposed by Krückel (2017, p. 62). Krückel transfers the figure of the hacker into the media pedagogical discourse. Hackers, as "those computer science specialists who misappropriated and reinvented the use of computers" (Bardeau and Danet 2012,

p. 9), stand for the subversive subversion of computer technology. As a subculture, the Hacker emerged "in the milieu of the 1960s protest movements" (ibid.). Chatfield (2013) points to the subversive connotation that accompanies the word hacker to the present day:

> The word has a deeper meaning in English, even outside the digital world, which has survived to this day: A hacker is someone who also concocts unconventional ways to use some system, deviating from the established rules when necessary and being something like "playfully clever". For some, this is still the true meaning of the word today (Chatfield 2013, p. 96 f.).

From this perspective, it seems obvious to make a form of 'intellectual hacking' usable as a figure of thought for the subversive epistemological attitude of postmodern epistemology. In the context of media Bildung, the figure of the cultural hacker formulated by Krückel condenses as a form of postmodern epistemological critique. Krückel understands the "cultural hacker [...] as a critical observer of the world" (Krückel 2017, p. 62). Via critical-analytical reflection, the cultural hacker can "read" or analytically objectify the cultural code of social practices. In terms of postmodern resistance, the cultural hacker is defined via a "critical in-world-ness" (Krückel 2017, p. 60) that is "linked to rebelling against the programming condition of society." (ibid.). This rebellion "represents the moment of disruption of order, that is, of the ordered, in a control society" (Krückel 2017, p. 60 f.). Krückel sees in the figure of the cultural hacker the possibility of opening "a space of thought" for "a scientific fiction" that "takes on the struggle in a digitally networked world to strengthen the role of educational sciences" (Krückel 2017, p. 62). Through postmodern epistemological critique, the cultural hacker provides a counterpoint to truth claims of metanarratives. In this context, the figure of the cultural hacker is to be thought in the plural: a critical epistemological attitude is realized in the course of collaborative knowledge production:

> Only in groups of authors, that is, intersubjectively, can postmodern man become the manipulator of his lifeworld [...] It is a dialogical, democratic as well as an artistic attempt at a utopian negotiation of the space of possibility that elevates telematic man to the manipulator and hacker of his lifeworld. It represents a turn against concepts that bind man in order and immaturity (Krückel 2017, p. 61.).

These processes of collaborative knowledge production can be realized through the use of digital media. Media Bildung takes place *in* and *through a* critical approach to media, which leads to a positively connoted relationship between the self and the world:

As a process, Bildung takes place in media and at the same time generates media. On the one hand, media are the background of Bildung as a process, but on the other hand also the result of Bildung processes. Suppose one wants to think of this simultaneously and therefore also grasp it theoretically. In that case, one must understand the medial articulation as a reflexive, processual Bildung in which the medium is co-generated (Meder 2011, p. 72).

Against the background of these considerations, postmodern media Bildung can be understood as a socially feedback process of perceiving media, dealing with media, and critically reflecting on media and the media-infrastructure of the world. Explorative curiosity and self-efficacy expectations characterize this process of perception, interaction, and critical reflection. Critical reflection constitutes an action- and production-oriented collective authorship in the sphere of digital media. In collaborative media action, the subject brings itself forth in the social context of media Bildung. Jörissen (2011) points to this link between mediality and subject in the socially contextualized process of media Bildung:

Decisive for the idea of media Bildung is, on the one hand, the fact that *articulations* cannot be separated from *mediality*, and, on the other hand, the fact that media spaces increasingly represent places of social encounter, that is, *media social arenas in the new media are* assuming ever greater importance for Bildung and subjectification processes. From the perspective of media Bildung, it is therefore essential to analytically recognize the reflexive potential of media spaces on the one hand and media forms of articulation on the other and to assess their educational value (Jörissen 2011, p. 225, editor's note).

In the course of collaborative processes of collective authorship, it is also essential to investigate the question of how digital media prefigure and determine construction processes of self/world relations:

What are the interpellative implications and subjectivizing dimensions of media? In Humboldt's sense, where is the potential for freedom through digital media so that the subject can develop 'powerfully'?

Following Baudrillard's analysis of the link between media use and power and relations, it can be concluded that media Bildung requires a dialogical use of (digital) media due to its emancipative implications. Against the background of these considerations on media Bildung, the relation between media Bildung and media literacy can also be defined: media competencies are empirical manifestations in dealing with media. They manifest themselves in skills and abilities to use media to establish, strengthen, and expand a positively connoted relationship to the self/

world *through* and *with media* use. "Media Bildung aims [...] at the *contexts for controlling media learning conditions* and at the organized use of media competencies as learning tools when dealing with demands of relevant environments" (Spanhel 2014, p. 124, e.i.o.). It should be added to this that media literacy is not a set of skills that can be defined conclusively. Instead, media literacy is a performative process in which an engagement with the media-structured world is constituted based on explorative curiosity and self-efficacy expectations. In the course of this, a critical and reflective approach to media emerges.

What do I want to do with the media?
Into which normative space of possibilities do media force me? or
What is the potential for interpellation prefigured by the media?
Or explicitly power-critical in the sense of Baudrillard: What relations of domination am I subjected to through media, how can I communicate in a self-determined and dialogical way despite these relations of domination?

The action and production orientation of such an understanding of media Bildung, which at the same time demands a postmodern reflexive 'agility,' corresponds to the participatory and ephemeral structure of the digital cultures. From this perspective, postmodern media Bildung *in* and *through* digital media can be understood as a subject-philosophically founded variety of digital cultures.

The understanding of media Bildung outlined here unfolds in Bildung learning. On the level of explorative learning, media Bildung unfolds in the university as emancipative Bildung space. In this process, media Bildung and the university as Bildung space prove to be capable of connecting to a tradition in which the university functions as a driver of innovation. In the transition from the electronic age to the digital age, the university has a prominent significance: technical innovations were developed, the emergence of hacker culture formed a technological subversion. In addition to that, the university represents an unconditional place of questioning and thus enables a critical examination of digitization processes. The potential of critical reflection can unfold in an unconditional university *in* and *through* media Bildung. The potential of media Bildung is discursively articulated in the context of the discussion of the 'digital humanities'/the location of the humanities in the digital age. Thus

in a world where the university is increasingly becoming just another knowledge-based organization driven by efficiency and flexibility, the Digital Humanities could become the ideal context in which the university is once again assigned the role of a public sphere. Questioning instrumentality is an essential step towards challenging the notion of knowledge as a commodity (Fabretti 2014, p. 101).

As a space of reflexive resistance and postmodern media Bildung, the university can be understood as a discursive figure that points to the possibilities of an Bildung based approach to digital media and to the processes of social transformation in the course of digitalization. A look at the University of Stanford, on the other hand, shows the extent to which critical thinking, digital development, and neoliberal restructuring of the university merge:

1969, a mainframe computer at Stanford Research Institute received the first message from a mainframe computer at the University of California, 600 km away.
In 1996, the Google precursor BackRub was developed by Larry Page and Sergey Brin at Stanford University.
In 1998, Derrida gave the lecture at Stanford University that evolved into the text "The Unconditional University."
Stanford, which as a research university is also committed to basic research, exerted a lasting influence on the cultural and technical developments that emerged in Silicon Valley. Neoliberal narrative topoi increasingly influence this meaning of Stanford.

As a critical examination of these neoliberal expressions of Stanford University, Thompson (2013) asks in the New Yorker magazine, "Is Stanford still a university?" (Thompson 2013, para. 1). Instead of educating students, they are dropping out to found start-ups with former students' professors on their boards. Thus, encouraged by the professors, more than a dozen students have left the university

> to work on a new technology start-up called Clinkle. Faculty members have invested, the former dean of Stanford's business school is on the board, and one computer-science professor who taught several of the employees now owns shares. The founder of Clinkle was an undergraduate advisee of the president of the university, John Hennessy, who has also been advising the company. Clinkle deals with mobile payments, and, if all goes well, there will be many payments to many people on campus. Maybe, as it did with Google, Stanford will get stock grants. There are conflicts of interest here; and questions of power dynamics. The leadership of a university has encouraged an endeavor in which students drop out to do something that will enrich the faculty (Thompson 2013, para. 1).

This process exemplifies how neoliberal logic displaces educational logic – also in the university, which loses its status as an unconditional institution (cf. Schimank 2014, p. 35). With the associated "turning away from the Humboldtian university idea" (ibid.), "efficiency and effectiveness enhancement of teaching and research" (ibid.) are to be produced. As a consequence, science and business become blurred, as is the case with the start-up Clinkle.

The tension between the ideal of a university as a postmodern Bildung space and neoliberal approaches, which can be paradigmatically demonstrated at Stanford University, shapes digital cultures. The common denominator of different digital cultures is – apart from the ephemeral structure – the activity/productivity of the individual. Digital cultures exhibit a range that extends from the pole of postmodern subversive diversity to the pole of neoliberal subjectivation. To gain a constructive approach to this tension, the approach of postmodern media Bildung can provide orientation points for identifying and experiencing the emancipative possibilities that the digital cultures bring with them...

# Literature

Adorno, T. W., & Horkheimer, M. (1997). *Dialektik der Aufklärung*. Frankfurt am Main: Suhrkamp.

Andrejevic, M. (2005). The Work of Watching One Another: Lateral Surveillance, Risk, and Governance. *Surveillance & Society* 2(4), 479–497.

Althusser, L. (1977). Ideologie und ideologische Staatsapparate. Aufsätze zur marxistischen Theorie. Hamburg: Vsa

Angermüller, J. (2013) Postmoderne. Zwischen Repräsentationskrise und Entdifferenzierung. In S. Moebius & A. Reckwitz (Hrsg.), *Poststrukturalistische Sozialwissenschaften* (S. 245–260). Frankfurt am Main: Suhrkamp.

Aristoteles (1995). Philosophische Schriften 3. Nikomachische Ethik. Hamburg: Felix Meiner.

Assange, J. (2013). The Banality of Don't be Evil. URL: http://www.nytimes.com/2013/06/02/opinion/sunday/the-banality-of-googles-dont-be-evil.html. Zuletzt zugegriffen: 13. Oktober 2017.

Assange, J. (2014). *When Wikileaks Meets Google*. New York: Or Books.

Assange, J, Appelbaum, J, Müller-Maghun, & Zimmermann, A. (2012). *Cypherpunks. Freedom and the Future of the Internet*. New York: Or Books.

Bak, M. P., & Kessler, T. (2012). Mir gefällt's, wenn's euch gefällt! Konformitätseffekte beiFacebook. *Journal of Business and Media Psychology* 3(2), 23–30.

Bandura, A. (1977). Self-efficacy: Toward a Unifying Theory of Behavioral Change. *Psychological Review* 84(2), 191–215.

Bardeau, F., & Danet, N. (2012). *Anonymous. Von der Spaßbewegung zur Medienguerilla*. Münster: Unrast.

Barlow, J. P. (1996). A Declaration of the Independence of Cyberspace URL: https://www.eff.org/de/cyberspace-independence. 23. August 2017.

Barthes, R. (2016). Der Tod des Autors. In F. Jannidis, G. Lauer, M. Martinez & S. Winko (Hrsg), *Texte zur Theorie der Autorschaft* (S. 185–193). Stuttgart: Reclam.

© Springer Fachmedien Wiesbaden GmbH, part of Springer Nature 2023
D. Kergel, *Digital Cultures*, https://doi.org/10.1007/978-3-658-35250-9

Barz, A. (2011). Zu den Begriffen Moderne und Postmoderne oder die Architektur als Zeitmaschine. Ein Versuch der Begriffsnäherung aus kunstwissenschaftlicher Perspektive. Denkmalpflege 3. http://edoc.hu-berlin.de/kunsttexte/2011-3/barz-andreas-2/PDF/barz. pdf. Zuletzt zugegriffen: 20. August 2017.

Baudrillard, J. (1978). Kool Killer oder der Aufstand der Zeichen. Berlin: Merve.

Baum, P. (2010). Große Erzählungen. In P. Baum & S. Höltgen (Hrsg.), Lexikon der Postmoderne. Von Abjekt bis Zizek (S. 87–88). Bochum: Projekt.

Baumann, Z. (1995). Moderne und Ambivalenz. Frankfurt am Main.

Ben Mhenni, L. (2011). Vernetzt Euch! Berlin: Ullstein.

Berner-Lee, T, Cailliau, R., Luotonen, A., Nielsen, H. F., & Secret, A. (1994). The World Wide Web. URL: http://storm.usc.edu/~Black/IML-400/fall-2012/readings/berners-lee_ et_al_the_world-wide_web.pdf. Zuletzt zugegriffen: 14. Oktober 2017.

Berry, D. (2014). Die Computerwende. Gedanken zu den Digital Humanities. In R. Reichert (Hrsg.), Big Data. Analysen zum digitalen Wandel von Wissen, Macht und Ökonomie (S. 47–64): Bielefeld: Transcript.

Bhabha, H. K. (2011). Die Verortung der Kultur. Tübingen: Stauffenberg.

Biebricher, T. (2012). Neoliberalismus zur Einführung. Hamburg: Junius.

Biermann, R., & Verständig, D. (2017). Das Netz im Spannungsfeld von Freiheit und Kontrolle. Ein kurzer Problemaufriss. In R. Biermann & D. Verständig (Hrsg.), Das umkämpfte Netz. Macht- und medienbildungstheoretische Analysen zum Digitalen (S. 1–15). Wiesbaden: VS Springer.

Bodeman, M. Y. (2011). Von Berlin nach Chicago und weiter. Georg Simmel und die Reise seines ,Fremden'. In H. A. Mieg, A. O. Sundsboe & M. Bieniok (Hrsg.), Georg Simmel und die aktuelle Stadtforschung (S. 185–213). Wiesbaden: Springer.

Bodó, B. (2011) You Have No Sovereignty Where We Gather – Wikileaks and Freedom, Autonomy and Sovereignty in the Cloud. doi: https://doi.org/10.2139/ssrn.1780519.

Boltanski, L., & Chiapello, È. (2013). Der neue Geist des Kapitalismus. Konstanz: Uvk.

Bourdieu, P. (1983). Die feinen Unterschiede. Frankfurt am Main: Suhrkamp.

Bourdieu, P. (1998). Gegenfeuer. Konstanz: Uvk.

Bourdieu, P. (1999). Die Regeln der Kunst. Genese und Struktur des literarischen Feldes. Frankfurt am Main: Suhrkamp.

Bourdieu, P. (2015). Die verborgenen Mechanismen der Macht. Schriften zu Politik & Kultur 1. Hamburg: Vsa.

Bourdieu, P. (2016), Neoliberalismus und neue Formen der Herrschaft. In Social Transformations. Resreach on Precarisation and Diversity – an interdisciplinary Journal 1(1). Art. 1.

Brandt, D. (2009). Postmoderne Wissensorganisation oder: Wie subversiv ist Wikipedia? Liberas. Library Ideas 14, 4–18.

Braun, K.-H. (1979). Kritik des Freudomarxismus. Köln: Pahl-Rugenstein.

Bröckling, U. (2013). Das unternehmerische Selbst. Soziologie einer Subjektivierungsform. Frankfurt am Main: Suhrkamp.

Bubner, R., & Hörisch, J. (1987). Zur dialektischen Bedeutung romantischer Ironie. In E. Behler (Hrsg.), Die Aktualität der Frühromantik (S. 85–95). Paderborn: Schöningh.

Buchanan, J., & Tullock, G. (1962). The Calculus of Consent: The Logical Foundations of Constitutional Democracy. Michigan: University of Michigan Press.

Buchanan, M. (2011). Privacy and Power in Social Space: Facebook. Stirling: University of Stirling: URL: https://dspace.stir.ac.uk/bitstream/1893/9150/1/PhD%20thesis%20.docx%20final%202%20Buchanan.pdf. Zuletzt zugegriffen: 23. September 2017.

Buckow, W.-D. (2008). Plädoyer für eine Neubestimmung von kulturellen Diskursen innerhalb der postmodernen Entwicklung. In S. Neubert, H.-J. Roth & E. Yildiz (Hrsg,), *Multikulturalität in der Diskussion, Neuere Beiträge zu einem umstrittenen Konzept* (S. 123–247). Wiesbaden: VS Springer.

Bush, V. (1945). As we may think. *The Atlantic Monthly* July 1945, 112–124.

Butler, J. (1991). *Das Unbehagen der Geschlechter*. Frankfurt am Main: Suhrkamp.

Butler, J (1995). Körper von Gewicht: die diskursiven Grenzen des Geschlechts Berlin: Suhrkamp

Butler, J. (1997). *The psychic life power. Theories in Subjection*. Stanford: Stanford University Press.

Butterwege, C., Lösch, B., & Ptak, R. (2008). *Kritik des Neoliberalismus*. Wiesbaden: VS Springer.

Camus, A. (1999). *Der Mythos von Sisyphos*. Reinbek bei Hamburg: Rowolth.

Castells, M. (2004). Der Aufstieg der Netzwergesellschaft. Das Informationszeitalter I. Opladen: Leske + Budrich.

Castells, M. (2005). Die Internet-Galaxie. Internet, Wirtschaft und Gesellschaft. Wiesbaden: VS Springer.

Castro Varela, M. d. M., & Dhawan, N. (2015). *Postkoloniale Theorie. Eine kritische Einführung*. Bielefeld: Transcript.

Careerbuilder.com (2012) Thirty-seven percent of companies use social networks to research potential job candidates. URL: http://www.careerbuilder.com/share/aboutus/pressreleasesdetail.aspx?id=pr691&sd=4/18/2012&ed=4/18/2099&siteid=cbpr&sc_cmp1=cb_pr691_. Zuletzt zugegriffen: 14. Oktober 2017.

Chatfield, T. (2013). *Digitale Kultur. 50 Schlüsselideen*. Heidelberg: Springer.

Chaos Computer Club (n.d.). Chaos Computer Club. URL: https://www.ccc.de/de/club. Zuletzt zugegriffen: 13. Oktober 2017.

Chiang, J. K. H., & Suen, H. Y. (2015). Self-presentation and hiring recommendations in online communities: Lessons from Linked. *Computers in Human Behavior* 48, 516–524.

Clifford, J. (1986). Introduction. Partial Truths. In Clifford, James & Marcus, George E. (Hrsg.), *Writing Culture. The Poetics and Politics of Ethnography* (S. 1–26). Berkley: University of California Press.

Coole, D. (2000). Negativity and Politics: Dionysus and Dialectics from Kant to Poststructuralism. London: Verso.

Culler, J. (2002). *Literaturtheorie. Eine kurze Einführung*. Stuttgart: Reclam.

Davenport, A. A. (2006). *Descartes's theory of action*. Leiden: Brill.

Deleuze, G. (2004). *Desert Islands: and other Texts*. Los Angeles: Semiotext(e).

Deleuze, G. (2005). Postskriptum über die Kontrollgesellschaften. In H. Breit, M. Rittberger & M. Sertl (Hrsg.), *Kontrollgesellschaft und Schule* (S. 7–14). Insbruck: Studien.

Deleuze, G., & Foucault, M. (1977). Die Intellektuellen und die Macht. In Foucault, M., *Der Faden ist gerissen* (S. 86–99). Berlin: Merve.

Deleuze, G., & Guattari, F. (1992). *Tausend Plateaus. Schizophrenie und Kapitalismus*. Berlin: Merve.

Derrida, J. (2015). *Die unbedingte Universität*. Frankfurt am Main: Suhrkamp. Alen

Descartes, R. (1972). Meditationen mit sämtlichen Einwänden und Erwiderungen. Hamburg: Meiner.

Dilthey, W. (1962). *Gesammelte Schriften Bd. 1*. Teubner: Stuttgart.

DIVSI – Deutsches Institut für Vertrauen und Sicherheit im Internet (2012): Milieu Studie zu Vertrauen und Sicherheit im Internet. Hamburg. https://www.divsi.de/wp-content/uploads/2013/07/DIVSI-Milieu-Studie_Gesamtfassung.pdf. Zugegriffen: 24. August 2017.

DIVSI – Deutsches Institut für Vertrauen und Sicherheit im Internet (2016): DIVSI Internet-Milieus 2016. Die digitalisierte Gesellschaft in Bewegung. URL: https://www.divsi.de/wp-content/uploads/2016/06/DIVSI-Internet-Milieus-2016.pdf. Zuletzt aufgerufen: 22 Mai 2017.

Doderer von, H. (1972). Grundlagen und Funktion des Romans. In H. Steinecke (Hrsg.), *Theorie und Technik des Romans im 20. Jahrhunderts* (S. 83–91). Tübingen: Max Niemeyer.

Dommann, M. (2008). Papierstau und Informationsfluss. Die Normierung der Bibliothekskopie. *Historische Anthropologie* 16(1), 31–54.

Downes, S. (2005). E-Learning 2.0. *e-learn-magazine*, URL: www.elearnmag.org/sub-page.cfm?section=articles&article=29-1. Zuletzt zugegriffen: 01. Juli 2017.

Duttweiler S., & Passoth, J.-H. (2016). Self-Tracking als optimierungsprojekt. In S. Duttweiler, R. Gugutzer, J.-H. Passoth & J. Strübing (Hrsg.), *Leben nach Zahlen. Self-Tracking als optimierungsprojekt* (S. 9–42). Bielefeld: Transcript.

Eco, U. (2003). *Nachschrift zum 'Namen der Rose'*. München: Dtv.

Ehrenberg, A. (2015). Das erschöpfte Selbst. Depression und Gesellschaft in der Gegenwart. Frankfurt am Main: Campus.

Engell, L. (2012). *Fernsehtheorie zur Einführung*. Hamburg: Junius.

El Difraoui, A. (2011). Die Rolle der neuen Medien im Arabischen Frühling. *Bundeszentrale für politische Bildung* 3. URL: http://www.bpb.de/internationales/afrika/arabischer-fruehling/52420/die-rolle-der-neuen-medien?p=all. Zuletzt zugegriffen: 23. Oktober 2017.

Fabretti, F. (2014), Eine neue Betrachtung der Digital Humanities. In R. Reichert (Hrsg.), *Big Data. Analysen zum digitalen Wandel von Wissen, Macht und Ökonomie* (S. 85–101). Bielefeld: Transcript.

Faucher, K. (2013). Thumbstruck: The Semiotics of Liking via the "Phaticon,". *Semiotic Review* Nr. 3. URL: http://www.semioticreview.com/pdf/open2013/faucher_semioticsofliking.pdf. Zuletzt zugegriffen: 23. September 2017.

Feierabend, S., Plankenhorn, T. & Rathgeb, T. (2014). JIM 2014 Jugend, Information, (Multi-) Media. Basisstudie zum Medienumgang. URL: http://www.sainetz.at/dokumente/JIM-Studie_2014.pdf. Zuletzt Zugegriffen am 26. Juli 2017.

Fichte, J. G. (1971). *Werke Bd1*. Hamburg: de Gruyter.

Firer-Blaess, S. (2016). The Collective Identity of Anonymous. Web of Meanings in a Digitally Enabled Movement. Uppsala: Uppsala University.

Fisch, S. (2015). Geschichte der Europäischen Universität. Von Bologna nach Bologna. München: Beck.

Fischer, J. (2006). Das Medium ist der Bote. Zur Soziologie der massenmedien aus der perspektive einer Sozialtheorie des Dritten. In A. Ziemann (Hrsg.), *Medien der Gesellschaft – Gesellschaft der Medien* (S. 21–41). Konstanz: Uvk.

Foucault, M. (1974). *Die Ordnung der Dinge*. Frankfurt am Main: Suhrkamp.

Foucault, M. (1997). *Überwachen und Strafen*. Frankfurt am Main.

Foucault, M (1989). *Der Gebrauch der Lüste*. Frankfurt am Main: Suhrkamp.

Foucault, M. (2016). Was ist ein Autor? In F. Jannidis, G. Lauer, M. Martinez & S. Winko (Hrsg), *Texte zur Theorie der Autorschaft* (S. 198–229). Stuttgart: Reclam.

Frank, M. (1984). *Was ist Neostrukturalismus? Vorlesungen*. Frankfurt am Main: Suhrkamp.

Freud, S. (1997). Vorlesungen zur Einführung in die Psychoanalyse und neue Folge. Frankfurt am Main: Fischer.

Friedrich, A., & Biermann, C. (2016). Digitale Begriffsgeschichte? Methodologische Überlegungen und exemplarische Versuche am Beispiel moderner Netzsemantik. URL: https://www.lt.informatik.tu-darmstadt.de/fileadmin/user_upload/Group_LangTech/publications/Friedrich_Biemann_Digitale_Begriffsgeschichte2016.pdf. Zuletzt zugegriffe. 16. Oktober 2017.

Galloway, A., & Thacker, E. (2014). Protokoll, Kontrolle und Netzwerke. In R. Reichert (Hrsg.), *Big Data. Analysen zum digitalen Wandel von Wissen, Macht und Ökonomie* (S. 290–311). Bielefeld: Transcript.

Georg-Lauer, J. (1988). ‚Das postmoderne Wissen‘ und die Dissens-Theorie von Jean-Francois Lyotard. In P. Kemper (Hrsg.), ‚*Postmoderne' oder der Kampf um die Zukunft* (S. 189–206). Frankfurt am Main: Fischer.

Gibson, E. J. (1998). Exploratory Behavior in the Development of Perceiving, Acting, and the Acquiring of Knowledge. *Annual Review of Psychology* 39(42), 417–430.

Giddens, A. (1976). Interpretative Soziologie. Eine kritische Einführung. Campus: Frankfurt.

Gilcher-Holtey, I. (2008). Die 68er Bewegung. Deutschland, Westeuropa, USA. München: Beck

Giroux, H. & Aronowitz, S. (1991). *Postmodern Education: Politics, Culture and Social Criticism*. Minneapolis: University of Minnesota Press

Gopnik, A. (2009). Kleine Philosophen. Was wir von unseren Kindern über Liebe, Wahrheit und den Sinn des Lebens lernen können. Berlin: Ullstein.

Grünewald, F., Mazandarani, E., Meinel, C., Teusner, R., Totschnig, M., & Willems, C. (2013). openHPI: Soziales und Praktisches Lernen im Kontext eines MOOC. In A. Breiter & C. Rensing (Hrsg.), *DeLFI 2013 – die 11. E-Learning Fachtagung Informatik* (S. 143–154). Bonn: Gesellschaft für Informatik.

Gugutzer, R. (2016). Self-Tracking als Objektivation des Zeitgeists. In S. Duttweiler, R. Gugutzer, J.-H. Passoth & J. Strübing (Hrsg.), *Leben nach Zahlen. Self-Tracking als Optimierungsobjekt* (S. 161–182). Bielefeld: Transcript.

Habermas, J. (1993). Der philosophische Diskurs der Moderne. Zwölf Vorlesungen. Frankfurt am Main.

Häcker, T (2011) „Portfolio revisited" – über Grenzen und Möglichkeiten eines viel versprechenden Konzepts. In T. Meyer, K. Mayberger, S. Münte-Goussar & C. Schwalbe (Hrsg.), *Kontrolle und Selbstkontrolle. Zur Ambivalenz von E-Portfolios in Bildungsprozessen* (S. 161–184). Wiesbaden: VS Springer.

Hahn, H. P. (2013). *Ethnologie. Eine Einführung*. Frankfurt am Main: Suhrkamp.

Han, B.-C. (2005). Hyperkulturalität. Kultur und Globalisierung. Berlin: Merve.

Han, B.-C. (2013). *Im Schwarm. Ansichten des Digitalen*. Berlin: Matthes & Seitz.

Hartmann, F. (2000). *Medienphilosophie*. Wien: Wuv.

Hartmann, F. (2006). Globale Medienkultur. Technik, Geschichte, Theorien. Wien: Wuv.

Hastedt, H. (2012). *Was ist Bildung. Eine Textanthologie.* Stuttgart: Reclam.

Hayek, F. A. v. (1981). *Recht, Gesetzgebung und Freiheit, 3.Bd.* München: Moderne Industrie.

Hebert, E. (2017). Machtstrukturen im Kontext von Überwachung im Internet und deren Relevanz für die Pädagogik. In R. Biermann & D. Verständig (Hrsg), *Das umkämpfte Netz. Macht- und medienbildungstheoretische Analysen zum Digitalen* (S. 67–79). Wiesbaden: VS Springer.

Heidegger, M. (2006). *Die Technik und die Kehre.* Pufflingen: Neske.

Hegel, G. F. W. (1952). *Phänomenologie des Geistes.* Hamburg: Meiner.

Hegel, G. F. W. (1976). *Ästhetik Bd.2.* Berlin: Aufbau Verlag.

Hepp, A. (2015). Überblicksartikel: Ethnizität und Transkulturalität. In A. Hepp, F. Krotz, S. Lingenberg & J. Wimmer (Hrsg.) (2015), *Handbuch Cultural Studies und Medienanalyse* (S. 299–304). Wiesbaden: VS Springer.

Herder, J. G. (2012). Auch eine Philosophie der Geschichte zur Bildung der Menschheit. Stuttgart: Reclam.

Herder, J. G. (2016). *Ideen zur Philosophie der Geschichte der Menschheit.* Altenmünster: Jürgen Beck.

Hessel, S. (2011). *Empört Euch!* Frankfurt am Main: Fischer.

Hill, S. (2017). Die Start-up Illusion. Wie die Internet-Ökonomie unseren Sozialstaat ruiniert. München: Knaur.

Hirsch, J. (2007). Die Universität: Elfenbeinturm, Wissensfabrik oder Ort kritischer Theoriebildung. In O. Brüchert & A. Wagner (Hrsg.), *Kritische Wissenschaft, Emanzipation und die Entwicklung der Hochschulen. Reproduktionsbedingungen und Perspektiven kritischer Theorie* (S. 241–248). Marburg. BdWi.

Hofmannsthal, H. (1951). *Gesammelte Werke in Einzelausgaben. Prosa II.* Frankfurt am Main: Fischer.

Hoinkis, T. (1997). Lektüre. Ironie. Erlebnis. System- und medientheoretische Analysen zur literarischen Ästhetik der Romantik. URL: http://www-brs.ub.ruhr-uni-bochum.de/neahtml/HSS/Diss/HoinkisTim/diss.pdf. Zuletzt zugegriffen: 16. Oktober 2017.

Holze, J. (2017). Das umkämpfte Wissen. Untersuchungen zu Aushandlungsprozessen in Wikipedia. R. Biermann & D. Verständig (Hrsg), *Das umkämpfte Netz. Macht- und medienbildungstheoretische Analysen zum Digitalen* (S. 95–110). Wiesbaden: VS Springer.

Howard, P. N., & Hussain, M. M. (2013). *Digital Media and the Arab Spring.* Oxford: Oxford University Press.

Humboldt, W. v. (1980). Theorie der Bildung des Menschen. In W. v. Humboldt, *Werke in fünf Bänden, Bd.1. Schriften zur Anthropologie und Geschichte* (S. 234–240). Stuttgart: Klett-Cotta.

Humboldt, W. v. (2010). Ideen zu einem Versuch, die Grenzen der Wirksamkeit des Staates zu bestimmen. Stuttgart: Reclam.

Iske, S. & Marotzki, W. (2010). Wikis: Reflexivität, Prozessualität und Partizipation. In B. Bachmair (Hrsg.), *Medienbildung in neuen Kulturräumen. Die deutschsprachige und britische Diskussion.* (S. 141–152). Wiesbaden: VS Springer.

Jörissen, B. (2011). „Medienbildung" – Begriffsverständnisse und Reichweiten. In H. Moser, P. Grell & H. Niesyto (Hrsg.), *Medienbildung und Medienkompetenz. Beiträge zu Schlüsselbegriffen der Medienpädagogik* (S. 211–235). München: Kopäd.

Jörissen, B. (2017). Digital/kulturelle Bildung. Plädoyer für eine Pädagogik der ästhetischen Reflexion digitaler Kultur. *OnlineZeitschrift Kunst Medien Bildung* | *zkmb 2017*. URL: http://zkmb.de/933. Zuletzt zugegriffen: 23. Oktober 2017.

Kabis, V. (2002), Weg mit der rosa Multikultibrille! Plädoyer für einen kulturalismuskritischen Ansatz in der interkulturellen Bildungsarbeit. URL: http://www.forum-interkultur. net/uploads/tx_textdb/16.pdf. Zuletzt zugegriffen: 30. Juli 2017.

Kalz, M., Specht, M., Klamma, R., Chatti, M. A. & Kober, R. (2007). Kompetenzentwicklung in Lernnetzwerken für das lebenslange Lernen. In U. Dittler, M. Kindt & C. Schwarz (Hrsg.), *Online-Communities als soziale Systeme. Wikis, Weblogs und Social Software im E-Learning* (S. 181–197). Münster: Waxmann.

Klamma, R., Chatti, M. A., & Kober, R. (2007). Kompetenzentwicklung in Lernnetzwerken für das lebenslange Lernen. In U. Dittler, M. Kindt & C. Schwarz (Hrsg.), *Online-Communities als soziale Systeme. Wikis, Weblogs und Social Software im E-Learning* (S. 181–197). Münster: Waxmann.

Kant, I. (1784). Beantwortung der Frage: Was ist Aufklärung? *Berlinische Monatsschrift* Nr. 12, 481–494.

Kant, I. (1956). *Kritik der reinen Vernunft*. Hamburg: Meiner.

Kantel, J. (2009). Per Anhalter durch das Mitmach-Web. Publizieren im Web 2.0: Von Social Networks über Weblogs und Wikis zum eigenen Internet-Fernsehsender. Heidelberg: Mitp.

Kammenhuber, N., Fessi, A., & Carle, G. (2010). Resilience: Widerstandsfähigkeit des Internets gegen Störungen–Stand der Forschung und Entwicklung. *Informatik-Spektrum* 33(2), 131–142.

Kammerl, R. (2017). Das Potential der Medien für die Bildung des Subjekts. Überlegungen zur Kritik der Subjektorientierung in der medienpädagogischen Theoriebildung. *Medien-Pädagogik* 27, 30–49. doi: https://doi.org/10.21240/mpaed/27/2017.01.14.X.

Kemmerling, A. (1996). *Ideen des Ich*. Frankfurt am Main: Suhrkamp.

Kergel, D. (2011a). *Subjektorientierte Sozialisationstheorie- und Praxis*. Aalborg: Institute for Learning and Philosophy, Aalborg University.

Kergel, D. (2011b). Two Aspects of Descartes, the poststructuralistic Career of the ‚I'. In J. Zeller & M. Rasmussen (Hrsg.), *Descartes som Filosof* (S. 34–48). Aalborg: Aalborg Universtetsforlag.

Kergel, D. (2013). Rebellisch aus erkenntnistheoretischem Prinzip. Möglichkeiten und Grenzen angewandter Erkenntnistheorie. Frankfurt am Main: Peter Lang.

Kergel, D. (2014). Forschendes Lernen 2.0 – Lerntheoretische Fundierung und Good Practice. In O. Zawacki-Richter, D. Kergel, P. Muckel, J. Stöter & K. Brinkmann (2014), *Offen für Neue Wege – Digitale Medien in der Hochschule* (S. 37–51) Münster: Waxmann.

Kergel, D. (2016). Bildungssoziologie und Prekaritätsforschung: Castingshows als Prekaritätsnarration. In R.-D. Hepp, R. Riesinger & D. Kergel (Hrsg.), *Precarity – Shift in the center of the Society. Interdiciplinary Perspectives* (S. 177–196). Wiesbaden: VS Springer.

Kergel, D. (2017a). Qualitative Bildungsforschung. Ein integrativer Ansatz. Wiesbaden: VS Springer.

Kergel, D. (2017b). The Postmodern Dialogue and the Ethics of Digital Based Learning. In D. Kergel, B. Heidkamp, P. Kjærsdam Telléus, T. Rachwal & Samuel Nowakowski

(Hrsg.), *The digital turn in Higher Education. Teaching and Learning in a changing World* (S. 47–56). Wiesbaden: Springer.

Kergel, D., & Heidkamp, B. (2015). Forschendes Lernen mit digitalen Medien. Ein Lehrbuch. #theorie #praxis #evaluation. Münster: Waxmann.

Kergel, D., & Heidkamp, B. (2016). Forschendes Lernen 2.0 Partizipatives Lernen zwischen Globalisierung und medialem Wandel. Wiesbaden: VS Springer.

Kerner, I. (2012). *Postkoloniale Theorien zur Einführung*. Hamburg: Junius.

Kiesel, D. (n.d.). Von der Ausländerpädagogik zur Interkulturellen Erziehung. Zur erziehungswissenschaftlichen Rezeption der Zuwanderung in die Bundesrepublik Deutschland: URL: http://www.lvr.de/media/wwwlvrde/jugend/beruns/politik_1/dokumente_53/2006 1129_11tejhk2vortragkiesel.pdf. Zuletzt zugegriffen: 14. Oktober 2017.

Kirpal, A., & Vogel, A. (2006). Neue Medien in einer vernetzten Gesellschaft: Zur Geschichte des Internets und des World Wide Web. *NTM Zeitschrift für Geschichte der Wissenschaften, Technik und Medizin* 14(3), 137–147.

Krämer-Badoni, T. (1978). Zur Legitimität der bürgerlichen Gesellschaft. Eine Untersuchung des Arbeitsbegriffs in den Theorien von Locke, Smith, Ricardo, Hegel und Marx. Frankfurt am Main: Campus.

Kristeva, J. (1990). *Fremde sind wir uns selbst*. Frankfurt am Main: Suhrkamp.

Krückel F. (2017). Bildung als projektive Einstellung in einer (Lebens-)Welt der Netzmetaphoriken. In R. Biermann & D. Verständig (Hrsg), *Das umkämpfte Netz. Macht- und medienbildungstheoretische Analysen zum Digitalen* (S. 51–66). Wiesbaden: VS Springer.

Laclau, E., & Mouffe, C. (2012). Hegemonie und radikale Demokratie. Zur Dekonstruktion des Marxismus. Wien: Passagen.

Lanier, J. (2010). Digitaler Maoismus. SZ.de, 10 Mai 2010. URL: http://www.sueddeutsche.de/kultur/das-so-genannte-web-digitaler-maoismus-1.434613. Zuletzt zugegriffen: 05. Oktober 2017.

Lehr, C. (2012). Web 2.0 in der universitären Lehre. Ein Handlungsrahmen für die Gestaltung technologiegestützter Lernszenarien. Boizenburg: Vwh.

Lessenich, S. (2012). *Theorien des Sozialstaats*. Hamburg: Junius.

Lessig, L. (2008). Remix, Making Art and Culture Thrive in the Hybrid Economy. London: Bloomsbury Academic.

Luhmann, N. (1995). Gesellschaftsstruktur und Semantik. Studien zur Wissenssoziologie der modernen Gesellschaft. Bd 4. Frankfurt am Main: Suhrkamp.

Lukács, G. (1971). Die Theorie des Romans. Ein gesichtsphilosophischer Versuch über die Formen der großen Epik. Neuwied: Luchterhand.

Lyon, D., & Baumann, Z. (2014). Daten, Drohnen, Disziplin. Ein Gespräch über flüchtige Überwachung. Frankfurt am Main: Suhrkamp.

Lyotard, J.-F. (1983). *The Postmodern Condition: A Report on Knowledge*. Manchester: Manchester University Press.

Marchart, O. (2013). Auf dem Weg in die Prekarisierungsgesellschaft. In O. Marchart (Hrsg.), Facetten der Prekarisierungsgesellschaft Prekäre Verhältnisse. Sozialwissenschaftliche Perspektiven auf die Prekarisierung von Arbeit und Leben (S. 7–20). Bielefeld: Transcript.

Marotzki, W. & Jörissen, B. (2008). Medienbildung. In U. Sander, F. v. Gross & K. U. Hugger (Hrsg.), *Handbuch Medienpädagogik* (S. 100–109). Wiesbaden: VS Springer.

Marotzki, W., & Jörissen, B. (2010) Dimensionen strukturaler Medienbildung. In B. Herzig, D. M. Meister, H. Moser & H. Niesyto (Hrsg), *Jahrbuch Medienpädagogik 8* (S. 19–39). Wiesbaden: VS Springer.

Martensen, M., Börgmann, K., & Bick, M. (2011). The Impact of Social Networking Sites on the Employer-Employee Relationship. In Proceedings of BLED Conference 2011. URL: http://aisel.aisnet.org/bled2011/54/. Zuletzt zugegriffen: 23. September 2017.

McLuhan, M. (1968). Die Gutenberg-Galaxis. Das Ende des Buchzeitalters. Düsseldorf: Econ.

Mecheril, P. (et al.) (2010). *Migrationspädagogik*. Weinheim: Beltz/Juventa.

Meder, N. (2011). Von der Theorie der Medienpädagogik zu einer Theorie der Medienbildung. J. Fromme, S. Iske & W. Marotzki (Hrsg), *Medialität und Realität. Zur konstitutiven Kraft der Medien*. (S. 67–81). Wiesbaden: VS Springer.

Mersch, D. (2006). *Medientheorien zur Einführung*. Hamburg. Junius.

Mishra, B. K. (2017). Digital Media in Resisting Social Inequality. The Indian Experience. In B. Heidkamp & D. Kergel (Hrsg.). *Precarity within the Digital Age. Media Change and Social Insecurity* (S. 123–133). Wiesbaden: VS Springer.

Mitrou, L., Kandias, M., Stavrou, V., & Gritzalis, D. (2014). Social Media Profiling: A Panopticon or omnipoticon tool? URL: https://www.infosec.aueb.gr/Publications/2014-SSN-Privacy%20Social%20Media.pdf. Zuletzt zugegriffen: 23. September 2017.

Mix, Y-G. (2010). Naturalismus, Fin die siècle, Expressionismus, 1890–1918. München: Dtv.

Moebius, S. (2016). Macht und Hegemonie. Grundrisse einer poststrukturalistischen Analytik der Macht. In S. Moebius & A. Rechwitz (Hrsg.), *Poststrukturalistische Sozialwissenschaften* (S. 158–173). Frankfurt am Main: Suhrkamp.

Mühsam, E. (1932). Die Befreiung der Gesellschaft vom Staat. Berlin: Fanal.

Münker, S. (2009). Emergenz digitaler Öffentlichkeiten. Die sozialen Medien im Web 2.0. Frankfurt am Main: Suhrkamp.

Musolff, H.-U. (1989). Bildung. Der klassische Begriff und sein Wandel in der Bildungsreform der sechziger Jahre. Weinheim: Deutscher Studienverlag.

Nadkarni, A., & Hofmann, S. G. (2012). Why do people use Facebook? *Personality and individual differences* 52(3), 243–249.

Nelson, T. H. (1965). Complex information processing: a file structure for the complex, the changing and the indeterminate. In *Proceedings of the 1965 20th national conference* (S. 84–100). New York: ACM.

Neubert, S., Roth, H.-J., & Yildiz, E. (2013). Multikulturalismus – ein umstrittenes Konzept. In S. Neubert, H.-J. Roth & E. Yildiz (Hrsg.), *Multikulturalität in der Diskussion Neuere Beiträge zu einem umstrittenen Konzept* (S. 9–29). Wiesbaden: VS Springer.

Niedermair, K. (1992). Das Ideal der philosophischen Postmoderne: Widerstand gegen die Okkupationen des Ideals in der Moderne. In A. Hütter, T. Hug, J. Perger (Hrsg.), *Paradigmenvielfalt und Wissensintegration. Beiträge zur Postmoderne im Umkreis von Jean-Francois Lyotard* (S. 87–98). Wien: Passagen.

Nietzsche, F. (1980). *Gesammelte Werke Bd.3*. Frankfurt am Main: Campus.

Nitsch, W. (2007). Strukturelle Bedingungen und Arbeitsformen kritischer Wissenschaft. In O. Brüchert & A. Wagner (Hrsg.), *Kritische Wissenschaft, Emanzipation und die Entwicklung der hochschulen. Reproduktionsbedingungen und Perspektiven kritischer Theorie* (S. 199–212). Marburg. BdWi.

Noetzel, (2008). Die Ironie der Politik. Der postmoderne Staat zwischen Komödie und Tragödie. In T. V. Winter & V. Mittendorf (Hrsg.), *Perspektiven der politischen Soziologie im Wandel von Gesellschaft und Staatlichkeit. Festschrift für Theo Schiller* (S. 39–47). Wiesbaden: VS Springer.

Ott, M. (2005). *Gilles Deleuze zur Einführung.* Hamburg: Junius.

Palm, G. (2006). CyberMedienWirklichkeit. Virtuelle Welterschließungen. Hannover: Heise.

Peitz, M., & Schwalbe, U. (2016). Zwischen Sozialromantik und Neoliberalismus – zur Ökonomie der Sharing-Economy. (No. 16-033). ZEW Discussion Papers.

Piaget, J. (1970). *Einführung in die genetische Erkenntnistheorie.* Frankfurt am Main: Suhrkamp.

Popper, K. (1973). *Logik der Forschung.* Tübingen: Mohr.

Prensky, M. (2001). Digital natives, digital immigrants part 1. *On the horizon* 9(5), 1–6.

Preyer, G. (n.d.). Modern und Postmodern im Kontext von Globalisierung. http://www.skopffm.de/Pagesneu/Glob.PDF. Zuletzt zugegriffen: 18. Juli 2017.

Rancière, J. (2008). Die Aufteilung des Sinnlichen. Die Politik der Kunst und ihre Paradoxien. Berlin: b_books.

Raulet, G. (1988). Leben wir im Jahrzehnt der Simulation? Neue Informationstechnologien und sozialer Wandel In P. Kemper (Hrsg.), *,Postmoderne' oder der Kampf um die Zukunft* (S. 165–188). Frankfurt am Main: Fischer.

Ravenscroft, A. (2011). Dialogue and connectivism: A new approach to understanding and promoting dialogue-rich networked learning. *The International Review of Research in Open and Distributed Learning* 12(3), 139–160.

Reichert, R. (2013). Die Macht der Vielen. Über den neuen Kult der digitalen Vernetzung. Bielefeld: Transcript.

Reichert, R. (2014). Einführung. In R. Reichert (Hrsg.), *Big Data. Analysen zum digitalen Wandel von Wissen, Macht und Ökonomie* (S. 9–31). Bielefeld: Transcript.

Reichert, R. (2016). Social Surveillance. Praktiken der digitalen Selbstvermessung in mobile Anwendungskulturen. In S. Duttweiler, R. Gugutzer, J.-H. Passoth & J. Strübing (Hrsg.), *Leben nach Zahlen. Self-Tracking als optimierungsprojekt* (S. 185–200). Bielefeld: Transcript.

Reckwitz, S. (2006). Die historische Transformation der Medien und die Geschichte des Subjekts. In A. Ziemann (Hrsg.), *Medien der Gesellschaft – Gesellschaft der Medien* (S. 89–107). Konstanz: Uvk.

Reese-Schäfer, W. (1989). *Lyotard zur Einführung.* Hamburg: Junius.

Reichardt, S. (2014). Authentizität und Gemeinschaft. Linksalternatives Leben in den siebziger Jahren. Frankfurt am Main: Suhrkamp.

Rieger-Ladich, M. (2012). Judith Butlers Rede von Subjektivierung. Kleine Fallstudie zur „Arbeit am Begriff". N. Ricken & N. Balzer (Hrsg.), *Judith Butler: Pädagogische Lektüren* (S. 57–74). Wiesbaden: VS Springer.

Ricken, N. (2006). Die Ordnung der Bildung. Beiträge zu einer Genealogie der Bildung. Wiesbaden: VS Springer.

Rohlfs, C., Harring, M., & Palentien, C. (2014). Bildung, Kompetenz, Kompetenz-Bildung. In C. Rohlfs, M. Harring & C. Palentien (Hrsg.), *Kompetenz-Bildung Soziale, emotionale und kommunikative Kompetenzen von Kindern und Jugendlichen* (S. 11–22). Wiesbaden: VS Springer.

Şahin, M. (2012). Pros and cons of connectivism as a learning theory. *International Journal of Physical and Social Sciences* 2(4), 437–454.

Said, E. (2009). *Orientalismus.* Frankfurt am Main: Fischer.

Schaupp, S. (2016). ‚Wir nennen es flexible Selbstkontrolle. Self-Tracking als Selbsttechnologie des kybernetischen Kapitalismus'. In S. Duttweiler, R. Gugutzer, J.-H. Passoth & J. Strübing (Hrsg.), *Leben nach Zahlen. Self-Tracking als optimierungsprojekt* (S. 63–86). Bielefeld: Transcript.

Schelhowe, H. (2007). Technologie, Imagination und Lernen: Grundlagen für Bildungsprozesse mit Digitalen Medien. Münster: Waxmann.

Schelling, F. W. J. (1964). Philosophische Untersuchungen. *Über das Wesen der menschlichen Freiheit.* Stuttgart: Leipzig.

Schimank, U. (2014). Krise – Umbau – Umbaukrise? Zur Lage der deutschen Universitäten. In N. Ricken, H.-C. Koller & E. Keiner (Hrsg.), *Die Idee der Universität – revisted* (S. 33–44). Wiesbaden: VS Springer.

Schlegel, F. (1980). *Literarische Notizen 1797–1801.* Frankfurt am Main: Ullstein.

Schmidt, E., & Cohen, J. (2013). The New Digital Age. Reshaping the Future of People, Nations and Business. London: John Murray.

Schneider, H. J. (2004). Anliegen und Ambivalenzen postmodernen Denkens. URL: http://www.forschungsnetzwerk.at/downloadpub/Schneider_potsdam_postmoderne_2004_beitrag.pdf. Zuletzt zugegriffen: 23. Juni 2017.

Schönberger, K. (2006). Wie falsche Informationen ‚wahre (‚Real Life'-) Ereignisse' schaffen. Persistente und rekombinante Formen aktivistischer Kommunikation durch Internet-Fakes. URL: http://www.code-flow.net/fake/book/schoenberger-dowethics-de.html. Zuletzt zugegriffen: 15. Oktober 2017.

Schulze, T. (2007). Modi komplexer und längerfristiger Lernprozesse. Beobachtungen und Überlegungen zu einer Theorie des Lernens. In H.-C. Koller, W. Marotzki & O. Sanders (2007). *Bildungsprozesse und Fremdheitserfahrungen. Beiträge zu einer Theorie transformatorischer Bildungsprozesse* (S. 141–159). Bielefeld: Transcript.

Schwalbe, C. (2011). Die Universität der Buchkultur im digital vernetzten Medium. In T. Meyer, W.-H. Tan, C. Schwalbe & R. Appelt (Hrsg.), *Medien & Bildung Institutionelle Kontexte und kultureller Wandel* (S. 179–192). Wiesbaden: VS Springer.

Seiler-Schiedt, E. (2013). Digitale Medien als Brücken zwischen Forschung und Lehre: Wie unterstützen Informations- und Kommunikationstechnologien die Forschungsuniversität? In C. Bremer & D. Krömker (Hrsg.), *E-Learning zwischen Vision und Alltag: zum Stand der Dinge* (S. 266–276). Münster: Waxmann.

Selke, H. (2016). Einleitung. In H. Selke (Hrsg.), Lifelogging. Digitale Selbstvermessung und Lebensprotokollierung zwischen disruptiber Technologie und kulturellem Wandel (S. 1–21). Wiesbaden: VS Springer.

Sera-Shriar, E. (2014). What is armchair anthropology? Observational practices in 19th-century British human sciences. *History of the Human Sciences* 27(2), 26–40.

Sesnik, W. (2014). Eine kritische Bildungstheorie der Medien In W. Marotzki & N. Meder (Hrsg.), *Perspektiven der Medienbildung* (S. 11–44). Wiesbaden: VS Springer.

Siemens, G. (2004). Connectivism. A learning theory for the digital age. *elearnspace.* URL: http://www.elearnspace.org/Articles/connectivism.htm. Zuletzt zugegriffen: 14. Oktober 2017.

Simmel, G. (1908). Exkurs über den Fremden. In G. Simmel, *Soziologie. Untersuchungen über die Formen der Vergesellschaftung*. Berlin: Duncker & Humblot. URL: https://userpages.uni-koblenz.de/~luetjen/sose14/sifre.pdf. Zuletzt zugegriffen: 14. Oktober 2017.

Spanhel, D. (2011). Medienkompetenz oder Medienbildung? Begriffliche Grundlagen für eine Theorie der Medienpädagogik. In H. Moser, P. Grell & H. Niesyto (Hrsg.), *Medienbildung und Medienkompetenz. Beiträge zur Schlüsselbegriffen der Medienpädagogik* (S. 95–120). München: Kopaed.

Spanhel, D. (2014). Der Prozess der Medienbildung auf der Grundlage von Entwicklung, Lernen und Erziehung. In W. Marotzki & N. Meder (Hrsg.), *Perspektiven der Medienbildung* (S. 121–148). Wiesbaden: VS Springer.

Specht, M., Kalz, M., & Börner, D. (2013). Innovation und Trends für Mobiles Lernen. In C. de Witt & A. Sieber (Hrsg.), *Mobile Learning. Potenziale, Einsatzszenarien und Perspektiven des Lernens mit mobilen Endgeräten* (S. 55–74). Wiesbaden: VS Springer.

Spivak, G. C. (2008). Can the Subaltern Speak? Postkolonialität und subalterne Artikulation. Wien: Turia + Kant.

Stalder, F. (2016). *Kultur der Digitalität*. Frankfurt am Main: Suhrkamp.

Standing, G. (2011). *The Precariat. The new dangerous Class*. London: Bloomsbury.

Steinert, H. (2007). Die Universität als Ort von Kritischer Theorie. In O. Brüchert & A. Wagner (Hrsg.), *Kritische Wissenschaft, Emanzipation und die Entwicklung der Hochschulen. Reproduktionsbedingungen und Perspektiven kritischer Theorie* (S. 17–27). Marburg. BdWi.

Stingelin, M. (2000). *Das Netzwerk von Gilles Deleuze. Immanenz im Internet und auf Video*. Berlin: Merve.

Strohschneider-Kohrs, I. (1967), Zur Poetik der deutschen Romantik II. Die romantische Ironie. In H. Steffen (Hrsg.), *Die Deutsche Romantik. Poetik, Formen und Motive* (S. 75–98). Göttingen: Vandenhoeck & Ruprecht.

Strübing, J., Kasper, B., & Staiger, L. (2016). Das Selbst der Selbstvermessung. Fiktion oder Kalkül? Eine pragmatische Betrachtung. In S. Duttweiler, R. Gugutzer, J.-H. Passoth & J. Strübing (Hrsg.), *Leben nach Zahlen. Self-Tracking als Optimierungsprojekt* (S. 271–291). Bielefeld: Transcript.

Suderland, M. (2014). ‚Worldmaking' oder ‚die Durchsetzung der legitimen Weltsicht'. Symbolische Herrschaft, symbolische Macht und symbolische Gewalt als Schlüsselkonzepte der Soziologie Pierre Bourdieus. In U. Bauer, U. W. Bittlingmayer, C. Keller & F. Schultheis (Hrsg.), *Bourdieu und die Frankfurter Schule. Kritische Gesellschaftstheorie im Zeitalter des Neoliberalismus* (S. 121–162). Bielefeld: Transcript

Taylor, E. B. (1871). Primitive Culture. Resreaches Researches into the Development of Mythology, Philosophy, Religion, Art, and Custo. London: John Murray.

Tenroth, H.-E. (2000). Geschichte der Erziehung. Einführung in die Grundzüge ihrer neuzeitlichen Entwicklung. München: Juventa.

The Mentor (2004). *The Hacker Manifesto. The Conscience of a Hacker*. URL: http://www.it-academy.cc/article/1375/Das+Manifest+von+The+Mentor.html. Zuletzt aufgerufen: 1. September 2017.

Thomas, A. (1993). Psychologie interkulturellen Lernens und Handelns. In A. Thomas (Hrsg.), *Kulturvergleichende Psychologie. Eine Einführung* (S. 377–424). Göttingen: Hogrefe.

Thompson, N. (2013). The End of Standford. URL: https://www.newyorker.com/tech/elements/the-end-of-stanford. Zuletzt zugegriffen: 26. September 2010.

Toffler, A. (1980). Die dritte Welle, Zukunftschance. Perspektiven für die Gesellschaft des 21. Jahrhunderts. München: Goldmann.
Tönnies, F. (2010). *Geist der Neuzeit*. München: Profil.
Turkle, S. (2011). *Life on Screen. Identity in the Age of Internet*. New York: Simon & Schuster.
Turner, V. W. (1998). Liminalität und Communitas. In A Belliger & D. J. Krieger (Hrsg.), *Ritualtheorien. Ein einführendes Handbuch* (S. 251–264). Opladen: Westdeutscher Verlag.
Vinnai, G. (2005). Utopie und Wirklichkeit der Universität (Abschiedsvorlesung): http://www.vinnai.de/utopie.pdf. Zugegriffen: 18. August 2017.
Vogel, M. (2013). 1968 als Kommunikationsereignis. Die Rolle des Fernsehens. In I. Gilcher-Holtey (Hrsg.), *Horizont-Verschiebungen des Politischen in den 1960er und 1970er Jahren* (S. 47–82). München: Oldenbourg.
Vormbusch. U. (2016). Taxonomien des Selbst. Zur Hervorbringung subjektbezogener Bewertungsordnungen im Kontext ökonomischer und kultureller Unsicherheit. In S. Duttweiler, R. Gugutzer, J.-H. Passoth & J. Strübing (Hrsg.), *Leben nach Zahlen. Self-Tracking als optimierungsprojekt* (S. 45–62). Bielefeld: Transcript.
Wandtke, A.-A. (2001). Copyright und virtueller Markt oder Das Verschwinden des Urhebers im Nebel der Postmoderne. Berlin: Humboldt Univeristät zu Berlin.
Weel v. d., A. (2011). *Changing our textual minds. Towards a Digital Order of Knowledge*. Manchester: Manchester University Press.
Weingart, P. (2003). *Wissenschaftssoziologie*. Bielefeld: Transcript
Welsch, W. (2010). Was ist eigentlich Transkulturalität? URL: http://www2.uni-jena.de/welsch/papers/W_Welsch_Was_ist_Transkulturalit%C3%A4t.pdf. Zuletzt zugegriffen: 15. Mai 2017.
Wild, R. (2017). Machttechnologien des Internets. In R. Biermann & D. Verständig (Hrsg.), *Das umkämpfte Netz. Macht- und medienbildungstheoretische Analysen zum Digitalen* (S. 81–94). Wiesbaden: VS Springer.
Willey, L., White, B. J., Domagalski, T., & Ford, J. C. (2012). Candidate-screening, information technology and the law: Social media considerations. *Issues in Information Systems* 13(1), 300–309.
Wittgenstein, L. (1963). Tractatus logico-philosophicus. Logisch-philosophische Abhandlung. Frankfurt am Main: Suhrkamp.
Wulf, C. (2007). Der Andere in der Liebe. In J. Bilstein & R. Uhle (Hrsg.). *Liebe. Zur Anthropologie einer Grundbedingung pädagogischen Handelns* (S. 35–48). Oberhausen: Athena.
Yaakoby, T. (2012). A Critical Examination of NeoMarxist and Postmodernist Theories as Applied to Education. Münster: Waxmann.
Yeh, S. (2013). Anything Goes? Postmoderne Medientheorien im Vergleich. Die (großen) (Medien-)Erzählungen von McLuhan, Baudrillard, Virilio, Kittler und Flusser. Bielefeld: Transcript.
Zantwijk, T. v. (2010). Wege des Bildungsbegriffs von Fichte zu Hegel. In J. Stolzenberg & L.-T. Ulrichs (Hrsg.), *Bildung als Kunst. Fichte, Schiller, Humboldt, Nietzsche* (S. 69–86). Berlin: de Gruyter.
Zima, P. V. (2007). *Theorie des Subjekts*. Tübingen: Francke.

Zimmerman, B. J. (1990). Self-Regulated Learning and Academic Achievement: An Overview. *Educational Psychologist* 25(1), 3–17.

Zippelius, R. (2003). *Geschichte der Staatsideen*. München: Beck.

Zizek, S. (1999). Das Unbehagen im Multikulturalismus. In. B. Kossek (Hrsg.), *Gegen-Rassismen: Konstruktionen – Interaktionen – Interventionen* (S. 151–166). Hamburg: Argument.

Zizek, S. (2011). *Lacan. Eine Einführung*. Frankfurt am Main: Fischer.

Zöller, G. (2010). „Manigfaltigkeit und Tätigkeit" Wilhelm von Humboldts kritische Kulturphilosophie. In J. Stolzenberg & L.-T. Ullrichs (Hrsg.). *Bildung als Kunst. Fichte, Schiller Humboldt, Nietzsche* (S. 171–183). Berlin: De Gruyter.

Zorn, I. (2011). Medienkompetenz und Medienbildung mit Fokus auf Digitale Medien. In: H. Moser, P. Grell & H. Niesyto (Hrsg.), *Medienbildung und Medienkompetenz. Beiträge zu Schlüsselbegriffen der Medienpädagogik* (S. 211–235). München: kopead.

Printed in the United States
by Baker & Taylor Publisher Services